Chemical Ecology

The Chemistry of Biotic Interaction

Thomas Eisner and Jerrold Meinwald
Editors
National Academy of Sciences

NATIONAL ACADEMY PRESS
Washington, D.C. 1995

NATIONAL ACADEMY PRESS · 2101 Constitution Avenue, N.W. · Washington, D.C. 20418

This volume is based on the National Academy of Sciences' colloquium entitled "Chemical Ecology: The Chemistry of Biotic Interaction." The articles appearing in these pages were contributed by speakers at the colloquium. Any opinions, findings, conclusions, or recommendations expressed in this volume are those of the authors and do not necessarily reflect the views of the National Academy of Sciences.

The National Academy of Sciences is a private, nonprofit, self-perpetuating society of distinguished scholars engaged in scientific and engineering research, dedicated to the furtherance of science and technology and to their use for the general welfare. Upon the authority of the charter granted to it by the Congress in 1863, the Academy has a mandate that requires it to advise the federal government on scientific and technical matters. Dr. Bruce M. Alberts is president of the National Academy of Sciences.

Library of Congress Cataloging-in-Publication Data

Chemical ecology : the chemistry of biotic interaction / Thomas Eisner and Jerrold Meinwald, editors.
p. cm.
Includes bibliographical references and index.
ISBN 0-309-05281-5 (alk. paper)
1. Chemical ecology. I. Eisner, Thomas, 1929– II. Meinwald, Jerrold, 1927– .
QH541. 15.C44C48 1995 95-18685
574.5—dc20 CIP

Printed in the United States of America

Table of Contents

Preface

Ours is a world of sights and sounds. We live by our eyes and ears and tend generally to be oblivious to the chemical happenings in our surrounds. Such happenings are ubiquitous. All organisms engender chemical signals, and all, in their respective ways, respond to the chemical emissions of others. The result is a vast communicative interplay, fundamental to the fabric of life. Organisms use chemicals to lure their mates, associate with symbionts, deter enemies, and fend off pathogens. Chemical ecology is the discipline that is opening our "eyes" to these interactions. It is a multifaceted discipline, intent on deciphering both the chemical structure and the information content of the mediating molecules. And it is a discipline in which discovery is still very much in order, for the interactions themselves remain in large measure to be uncovered.

Chemical ecology has made major progress in recent decades. This reflects, in part, the extraordinary technical innovation that has taken place in analytical chemistry. Highly improved procedures are now available for separating complex mixtures into their individual components, as well as for quantitating and chemically characterizing designated compounds. There has also been a vast increase in the sensitivity of the techniques. Where gram quantities were once needed for elucidation of chemical structure, milligram or even microgram quantities may now suffice. These refinements in sensitivity are of particular importance, given that organisms often produce their signal molecules in vanishingly small amounts.

Progress in chemical ecology has also been fostered by advances in biology itself. Chemical interactions in nature are often social, in the sense that they occur between conspecifics. Conceptual advances in behavioral biology, particularly sociobiology, have helped put new slants on inquiries into such social phenomena as mate attraction, sexual selection, parental investment, caste determination, and colony organization, all frequently mediated by chemicals. The questions themselves, answered at one level of organization, often lead to inquiries at another level. Studies of pheromones, for instance, first with insects and then with selected mammals, were doubtless influential in prompting the highly promising current inquiries into pheromonal communication in humans. Other biological disciplines are also proving relevant. Virtually every chemically mediated interspecific interaction, whether between predator and prey, herbivore and plant, or parasite and host, lends itself to interpretation in the broadest evolutionary, ecological, population-biological, and molecular-biological terms.

Molecular biology may, in fact, increasingly shape the questions that are asked in chemical ecology. How do given signal molecules arise in the course of evolution? How are they synthesized, and how is the rate and timing of their production controlled? How are they recognized at the level of the receptor? How do noxious chemical signals, designed to repel or poison, affect their intended targets? How is it that receiver organisms are sometimes able to circumvent, counteract, or even secondarily employ, such offensive chemicals? Molecules that transmit information between organisms are a fundamental part of the regulatory chemicals of nature. The rules that apply to intraorganismal chemical regulation apply in large measure to them as well.

Molecules that have signal value in nature sometimes prove to be of use to humans. One need only cite the example of medicinals to underscore the point. Major recent additions to our therapeutic arsenal include ivermectin, cyclosporin, FK-506, and taxol, compounds that can all be expected to have evolved as signaling agents. Many and varied benefits can be expected to be derived from an ongoing search for natural products. Chemical ecologists should become active participants in this search. They have the expertise, gained through laboratory and field experimentation and observation, to rate species by "chemical promise" and therefore to aid in the important task of selecting species for chemical screening. Chemical ecologists are also in a position to provide some assessment of the hidden value of nature. The search for natural products has essentially only begun. Most species, especially microbial forms and invertebrates, remain to be discovered, let alone to be screened for chemicals. What remains unknown is of immense potential value, and deserving of protection, lest we be forever impoverished by its loss. To help in the preservation effort, chemical ecologists will need to speak out as conservationists.

The essays that follow are synopses of lectures delivered at a colloquium on chemical ecology. Almost 150 participants attended the proceedings. The papers do not provide an overview of the discipline but rather give a glimpse into selected research areas that are contributing to advancement of the field. We are immensely grateful to our invited speakers, both for the quality of their communications and for the personal enthusiasm they brought to the meeting. Discussions were convivial and much enlivened by the youthfulness of most of the audience. Four participants, Ian T. Baldwin, Gunnar Bergström, Arnold Brossi, and Amos B. Smith III deserve special thanks, for presiding over the sessions and for leading the discussions. We are also grateful to Jack Halpern, Vice President of the Academy, for asking us to organize the colloquium, and to Bruce Alberts, President of the Academy, for providing introductory remarks at the meeting. For help in preparation of the colloquium we are indebted to Kenneth R. Fulton and Jean Marterre of the Academy and especially to our Cornell associates, Janis Strope and Johane Gervais.

Financial resources for the colloquium were provided by the Academy and supplemented by contributions from a number of industries (American Cyanamid Company, E. I. DuPont de Nemours & Co., Givaudan–Roure Corporation, Merck & Co., Monsanto, Rohm and Haas Company, Schering–Plough Research Institute, Sterling–Winthrop Inc., Syntex, Takasago International Corporation, Zeneca Inc.), to which we are much indebted.

Thomas Eisner
and Jerrold Meinwald

Chemical Ecology

The Chemistry of Defense: Theory and Practice

MAY R. BERENBAUM

Irrespective of taxon, the chemicals that play a prominent role in interspecific interactions are rarely the same substances used by an organism to meet the daily challenges of living, such as respiration, digestion, excretion, or, in the case of plants, photosynthesis. They are, in both plants and animals, of "a more secondary character" [to borrow a phrase from Czapek (1)]. These secondary compounds are generally derived from metabolites that do participate in primary physiological processes. In plants, for example, secondary compounds such as alkaloids, coumarins, cyanogenic glycosides, and glucosinolates derive from amino acids; tricarboxylic acid cycle constituents are involved in the formation of polyacetylenes and polyphenols; glucose, aliphatic acids and other "primordial molecules" (2) play a role not only in primary metabolism but in secondary metabolism as well. In insects, many defensive secretions are derived from the same amino acids used to construct proteins [among them, quinones in many beetles and cockroaches derive from tyrosine, formic acid in ants from serine, isobutyric acid in swallowtail caterpillars from isoleucine and valine, and alkyl sulfides in ants from methionine (3, 4)]. Presumably, secondary compounds are physiologically active in nonconspecific organisms precisely because of their secondary nature; it is to be expected that most organisms possess effective means for metabolizing, shunting around, or

May Berenbaum is professor and head of the Department of Entomology at the University of Illinois, Urbana-Champaign.

otherwise processing primary metabolites and it is the unusual chemical that circumvents these mechanisms to cause toxicity.

Unlike primary metabolites, which are practically universal constituents of cells, tissues, and organs, secondary compounds are generally idiosyncratic in distribution, both taxonomically and ontogenetically. Chlorophyll, for example, the principal photosynthetic pigment, is found in virtually all species of angiosperms, in virtually all life stages of virtually all individuals. In contrast, the furanocoumarins are secondary compounds known from only a handful of angiosperm families (5). Within a species (e.g., *Pastinaca sativa*), there is variability in furanocoumarin content and composition among populations (6, 7); within an individual, there is variation among body parts during any particular life stage (8) and temporal variation in the appearance of these compounds over the course of development (9); there are even differences in the content of individual seeds, depending upon their location in an umbel (10), fertilization history (11), and their position within the schizocarp (12).

Secondary chemicals are by definition taxonomically restricted in distribution, yet despite this fact there are patterns in production and allocation that transcend taxa (13). Their presence in an organism is generally characterized by specialized synthesis, transport, or storage. Levels of abundance are subject to environmental or developmental regulation and, unlike primary constituents, which may be present in virtually all cells of an organism, chemical defenses are typically compartmentalized, even in those cases in which the chemicals are acquired exogenously, as when sequestered from a food source. There often exists a system for external discharge, delivery, or activation, not only as a means of ensuring contact with a potential consumer but also as a means of avoiding autotoxicity until a confrontation arises; and of course these compounds are almost invariably, by virtue of structure, chemically reactive (e.g., able to be taken up by a living system, to interact with a receptor or molecular target, and to effect a change in the structure of the molecular target). The remarkable convergence of structural types in plant and insect secondary metabolites is at least suggestive that the processes leading to biological activity in both groups share certain fundamental similarities (14).

Secondary chemicals can be said to be defensive in function only if they protect their producers from the life-threatening activities of another organism. Distinguishing between offensive and defensive use of chemicals is difficult, and present terminology does little to assist in making that distinction. The term "allomone" is frequently used synonymously with "chemical defense," yet allomones are not necessarily defensive in function. An allomone has been defined as a chemical substance beneficial to its producer and detrimental to its recipient (15), so chemicals used by a predator to lure prey (16) are rightly regarded as allomonal but are not obviously defensive. By the same token, chemicals that reduce competition for limited resources, clearly beneficial to the producer, may be defensive of those resources but are not necessarily defensive in the life of the organism producing them. Allelopathic compounds produced by a plant species may increase fitness of that plant by preempting a resource, such as water or soil nitrogen, that might otherwise be exploited by other plants (17), but in the sense that such compounds can kill potential competitors (such as nonconspecific seedlings) they are used in an offensive fashion, as for range expansion at the expense of another organism.

A defensive chemical, then, is a substance produced in order to reduce the risk of bodily harm. As such, most are poisons—defined as "any agent which, introduced (especially in small amount) into an organism, may chemically produce an injurious or deadly effect" (18). This rather restrictive definition may not be universally embraced by chemical ecologists. On one hand, the definition implies an interaction with another organism and, particularly with respect to plants, secondary compounds may fulfill many functions in the life of the producer organism other than producing injurious or deadly effects on other organisms (19, 20). Many plant secondary compounds, for example, are inducible by UV light and presumably serve to protect (or "defend") plants from damaging effects of UV exposure (21); by no stretch of the imagination can such compounds be considered poisons, since they exert no injurious effects on the damaging agent, the sun. In this context, they can no more be considered "defenses" than cell wall constituents can be considered "defenses" against gravity. On the other hand, some investigators, while acknowledging the fact that secondary chemicals have deleterious effects on other organisms, are reluctant to ascribe their presence, particularly in plants, to selection pressure exerted by those organisms (22–24). Calling certain secondary chemicals "defenses" would be giving credence to the assertion that they exist only by virtue of the selection pressures exerted by consumers. Nonetheless, an examination of the distribution, pattern of allocation, chemical structure, and modes of action of secondary compounds in a broad cross section of organisms reveals so many striking convergences and similarities that the notion that variation in the distribution and abundance of chemicals that act as poisons results at least in part from selection by consumer organisms certainly seems tenable, if not inescapable.

DISTRIBUTION OF DEFENSES

One line of evidence, admittedly circumstantial, that consumers have influenced the evolution of chemical defenses is their taxonomic distribution. There are entire phyla in which chemical defenses have never been identified (Table 1). Undoubtedly, in many cases this absence of chemical defenses may result simply from an absence of studies explicitly designed to discover them—for many small, obscure organisms, life histories, let alone chemistries, are poorly known. This problem may not be as severe as it might appear, because chemically defended organisms often call attention to themselves by way of aposematic coloration (Table 1) (in fact, it may well be that effective defenses, particularly chemical ones, may be a prerequisite for a conspicuous life-style among smaller organisms). Nonetheless, any reported absence of chemical defense may be artifactual due to incomplete information. With that caveat in mind, it is interesting to note that conspicuously abundant on the list of the chemically defenseless are phyla comprised exclusively of parasitic animals. As well, chemical defenses are absent in entirely parasitic orders within classes (Phthiraptera and Siphonaptera in the class Insecta, for example). These organisms are subject to mortality almost exclusively by their hosts, and poisoning or otherwise severely impairing a host is unlikely to enhance lifetime fitness of a parasite (particularly those parasites that cannot survive more than a few hours without one).

Chemical defenses are also rare in organisms at the top of the food chain—organisms that are themselves at low risk of being consumed. Large vertebrates, by virtue of size, speed, and strength, often occupy that position in both terrestrial and aquatic ecosystems (carnivores and odontocetes, for example). Chemically defended mammals include skunks and the duck-billed platypus, both opportunistic scavengers (32). Among birds, chemical defense has been demonstrated to date only in the pitohui (25), which feeds on leaf litter invertebrates (J. Daly; ref. 79), but likely exists in the conspicuously colored female hoopoe, which "has a strongly repulsive musty smell that emanates from her preen gland, and is believed to have a protective function like attar of skunk" (33). Hoopoes are also opportunistic feeders that consume debris along with insects and other invertebrates. It is somewhat surprising that chemical defenses are not more frequently encountered among small birds, but the absence of reports may be due to the tendency of investigators to assume conspicuous plumage results from sexual selection, rather than aposematism and distastefulness (25).

In contrast with fast, strong predators, organisms with a limited range of movement, or limited control over their movements—those that cannot run away from potential predators—are well represented among the chemically defended (Table 1). Sessile marine invertebrates are particularly accomplished chemists; these include in their ranks sponges, anthozoan corals, crinoid echinoderms, polychaetes, bryozoans, brachiopods,

TABLE 1 General survey of the distribution of chemical defenses (3, 25–30) and certain life history characteristics (31) in animals

Animal phyla in which chemical defenses are rare or unreported

Platyhelminthes,*† 15,000 spp.	Phoronida,‡ 10 spp.	Chordata: Vertebrata
Rhynchocoela,† 650 spp.	Pogonophora,‡ 80 spp.	Chondrichthyes, 800 spp.
Gnathostomulida, 80 spp.	Onychophora, 70 spp.	Aves,† 8600 spp.
Gastrotricha, 400 spp.	Echiurida, 100 spp.	Mammalia, 4500 spp.
Rotifera, 2000 spp.	Tardigrada, 400 spp.	
Kinorhyncha, 100 spp.	Pentastomida,* 60 spp.	
Acanthocephala,* 500 spp.	Priapulida, 9 spp.	
Nematoda,* 1500 spp.	Chaetognatha, 65 spp.	
Nematomorpha,* 100 spp.	Chordata: Cephalochordata, 28 spp.	

Animal phyla in which autogenous chemical defenses are documented

- Porifera,†‡ 5000 spp.
  - Sesquiterpenes, sesterterpenes, dibromotyrosine derivatives, isonitriles, isothiocyanates, polyalkylated indoles, macrolides, quinones, ancepsenolides, sterols
- Coelenterata:†‡ Anthozoa
  - Alcyonaria, 9000 spp.
    - Sesquiterpenes, diterpenes, alkaloids, prostaglandins, pyridines
  - Zoantharia
    - Peptides, proteins
- Ectoprocta,‡
  - Tambjamine pyrroles, 4000 spp.
- Brachiopoda,‡ 300 spp.
- Mollusca:† Gastropoda, 50,000 spp.
  - Opistobranchia
    - Sesquiterpene dialdehydes, dimenoic acid glycosides, haloethers
  - Prosobranchia
    - Triterpenes
  - Pulmonata
    - Polyproprionates
- Annelida (Polychaeta†‡), 5300 spp.
  - Phenolics
- Arthropoda:† 800,000 spp.
  - Insecta
    - Hydrocarbons, alcohols, aldehydes, ketones, carboxylic acids, quinones, esters, lactones, phenolics, steroids, alkaloids, cyanogenic glycosides, sulfides, peptides, proteins
  - Arachnida
    - Quinones, alkaloids, cyanogenic glycosides
- Echinodermata,† 6000 spp.
  - Holothuroidea
    - Steroidal glycosides, saponins
  - Crinozoa‡
    - Polyketide sulfates
  - Asterozoa
    - Phenolics, saponins, steroidal glycosides
- Chordata, 1250 spp.
  - Tunicata†‡
    - Bipyrrole alkaloids, cyclic peptides, quinones, macrolides, polyethers
  - Vertebrata: Osteichthyes†, 22,000 spp.
    - Alkaloids, peptides
  - Amphibia,† 3150 spp.
    - Alkaloids
  - Reptilia,† 7000 spp.
    - Alkaloids, hydrocarbons, aldehydes, acids

*Many species parasitic.
†Many species conspicuously colored.
‡Many species sessile.

and tunicates (26, 34). Completely consistent with the pattern is the virtually universal presence of toxins in plants, ranging from mosses to angiosperms (4), most of which remain firmly rooted to the ground for most of their lives and occupy the bottom rung of most food chains. It is interesting to note that chemically defended taxa tend to be more speciose than those lacking chemical defenses, but whether this relationship reflects sampling vagaries or causation is anybody's guess.

PATTERNS OF ALLOCATION

Secondary chemistry differs from primary chemistry principally in its distributional variability and it is this variability that has intrigued ecologists for the past 30 years. Theories [or provisional hypotheses (35)] to account for the structural differentiation and function of secondary metabolites, as well as the differential allocation of energy and materials to defensive chemistry, abound, but they are almost exclusively derived from studies of plant–herbivore interactions (Table 2). This emphasis may be because the function of secondary chemicals in plants is less immediately apparent to humans, who have historically consumed a broad array of plants without ill effects, so alternative explanations of their presence readily come to mind. The fact that animals upon disturbance often squirt, dribble, spray, or otherwise release noxious substances at humans and cause pain leads to readier acceptance of a defensive function [although there are skeptics who are unconvinced of a

TABLE 2 Chemical defense theories

	Ref.
Theories to account for allocation of chemical defenses in plants	
Plant stress hypothesis	36
Latitudinal pest pressure gradients	37
Plant defense guilds	38
Apparency theory	39
Toxin/digestibility reduction continuum	40
Optimal defense theory	41
Optimal defense theory	42
Carbon/nutrient balance hypothesis	43
Resource availability hypothesis	44
Environmental constraint hypothesis	45
Plant vigor hypothesis	46
Growth-differentiation balance hypothesis	47
"Probability" hypothesis	23
Theories to account for allocation of chemical defenses in animals	
?????	

defensive function of certain animal secondary compounds—Portier (48), for example, reports that "Certains auteurs voient dans les glandes nucales (of swallowtail caterpillars) un appareil d'élimination de substances toxiques ou tout au moins inutilisables contenus dans la nourriture"]. Why plants, by virtue of their ability to photosynthesize, should occupy a unique place in theories of chemical defense allocation is unclear. Plants produce secondary compounds as derivatives of primary metabolism; animals do the same. In fact, plants may be rather unrepresentative of chemical defense strategies as a whole in that they rarely coopt defense compounds from other organisms via sequestration, although there are exceptions to the general rule [e.g., parasitic plants (49–51)].

The relative importance of consumer selection pressure in determining patterns of production of secondary compounds varies with the theory. Coley *et al.* (44) suggest that resource availability and the concomitant growth rate of a plant, more than its potential risk of herbivory or its historical association with herbivores, determine the type and quantity of chemical defenses in plants; while "the predictability of a plant in time and space may influence the degree of herbivore pressure . . . it should be included as a complementary factor," rather than as the sole driving force in the evolution of chemical defenses and their allocation patterns. Bryant *et al.* (43) suggest that carbon and nutrient availability alone can determine patterns of chemical defense allocation; according to this hypothesis, "environmental variations that cause changes in plant carbohydrate status will lead to parallel changes in levels of carbon-based secondary metabolites" (52). Such theories, along with the contention that "the evolution of plant defense may . . . have proceeded independent of consumer adaptation" (23), are in many ways reincarnations of "overflow metabolism" or "biosynthetic prodigality" hypotheses that reappear intermittently (53, 54). Yet how overflow metabolism can generate and maintain biochemical *diversity*—hundreds of biosynthetically distinct and unique classes of secondary metabolites—is an enigma.

Essentially (and rightly) undisputed is the fact that novel secondary compounds arise by genetic accident—by mutation or recombination—so it is not altogether surprising that, given the idiosyncratic nature of mutations, the distribution of biosynthetic classes of compounds is idiosyncratic as well. At issue, however, is how certain mutations become established within a population or species. Mutant individuals can increase in representation in populations either as a result of positive selection or as a result of random genetic events, such as drift; fixation by drift occurs only when there is no negative selection against the trait. It is a virtual certainty that at least some portion of the chemical variability of plants (and probably of all organisms) is nonadaptive—vestiges of past selection pressures no longer experienced due to extinctions, transient occurrences of secondary metabolites generated by indiscriminant enzymatic transformations, and the like (23). But predictable and highly specific accumulation of particular types of chemicals in taxonomically related species in particular organs in particular portions of the life cycle regulated by promoters that respond to chemical cues from consumer organisms (55, 56) seems inconsistent with such nonspecific processes.

Very little discussion to date in plant–insect interactions has centered on why certain secondary metabolites are built the way they are and why they act the way they do on consumer organisms. In vertebrates, "overflow metabolism" ends up almost exclusively converted to adipose tissue, despite the demonstrable ability of at least a few vertebrates to manufacture secondary metabolites; such tissue, in times of nutrient deprivation, is in fact readily mobilized by its producer. Why plants, which have the capacity to make glucose and, from glucose, the storage material starch, should make secondary compounds as "overflow" metabolites is unclear. Since glucose is a starting material for much of secondary metabolism, it is difficult to conceive of how such elaborate pathways could evolve in the absence of any selection pressure other than whatever problems may be associated with fixing too many carbon atoms. The suggestion that "high tissue carbohydrate carbon concentrations cause more carbon to flow through pathways leading to the synthesis of carbon-based secondary metabolites [and that] this effect of mass action on reaction rates is . . . stronger than any enzymatic effects such as feedback inhibition due to end-product accumulation" (52) runs counter to the many observations that secondary metabolite production is otherwise finely regulated by enzymes physically, temporally, and developmentally within a plant (57). In fact, such physiological responses to sunlight sound positively pathological.

Recent allocation theories generally classify chemicals based on criteria other than specific structure or biosynthetic origin. "Carbon-based" compounds (43, 52) are considered as a more-or-less homogeneous group, despite the fact that they include biosynthetically unrelated groups with wildly different activities as well as transport and storage requirements. The same is true for "N-based" compounds; the dichotomy between N-based and C-based compounds does not appear to take into consideration the fact that N-based compounds may actually contain more carbon atoms than smaller C-based compounds, and C-based compounds may require more investment in N-based enzymes for synthesis and storage than do many N-based compounds. Rather, there is a focus on whether or not secondary compounds are easily metabolized, or turned over, by the plants producing them [as in Coley *et al.* (44), "mobile" and "immobile" defenses] or whether they are accumulated in large or small quantities (39). Yet more defense does not necessarily lead to better defense—adding small amounts of biosynthetically different chemicals may by synergism enhance existing defenses more effectively than greatly increasing the concentration of those existing chemicals (58). Such interactions cannot be evaluated if only a single type of chemical is quantified. Secondary chemicals are not like muscles—pumping them up is not always the most effective strategy for overcoming an opponent.

Despite the growing number of studies failing to confirm at least one of the predictions of the carbon/nutrient balance hypothesis [24 such studies are cited in Herms and Mattson (59)], it remains popular as a testable hypothesis, as do several of its predecessors as well as its successors. In fact, none of these theories has really ever been resoundingly rejected; they all more-or-less coexist, by virtue of supportive evidence in some system or other. Studies of plant–herbivore interactions are in a sense unique in the field of chemical ecology; no other area is quite so rife with theory. One problem with attempting to develop an all-encompassing theory to account for patterns of defense allocations [as called for by Price (35) and Stamp (60)] is that such a theory may not be a biologically realistic expectation even for just the plant kingdom. Most theories certainly suit the specific systems from which they were drawn, which of necessity constitute only a tiny fraction of all possible types of interactions; it is when they are generalized that the fit breaks down. That many theories coexist is at least in part due to the fact that they are not mutually exclusive—they all share certain elements. If there is a recurrent theme in the past century of discussion, it is that chemical defenses confer a benefit and exact a cost. Much current controversy centers not on the existence of costs and benefits but on the magnitude of those values. Resource availability hypotheses (43, 44) focus on material costs of production; herbivory-based hypotheses (39–42) focus on benefits accrued. So if a ratio is to be tested, better that it simply be the benefit/cost ratio—that is, the benefits of a chemical, in terms of fitness enhancement in the presence of consumers, relative to the costs of a chemical, in terms of fitness decrements resulting from its production, transport, storage, or deployment.

This approach is not without its shortcomings, the greatest of which is that costs and benefits have proved to be exceedingly difficult to measure. There is far from a consensus on what constitutes an appropriate demonstration of costs of chemical defense (57, 61, 62). In many theoretical discussions of costs of defense, particularly in plants, costs are measured in terms of growth rates (44, 59, 63), rather than in terms of reproductive success. In many empirical estimates of costs, the chemical nature of the defense is not defined (64) or secondary metabolites are measured in bulk (65), without any regard to their individual activities.

On the other side of the coin, measuring the benefits of synthesizing a particular chemical compound requires an intimate knowledge of the interactions of the producer organism with its biotic environment. Bioassays of both plant and animal defenses tend to be done with isolated compounds against laboratory species that are easily reared (66); in the future, bioassays may need to be done with more ecologically appropriate species (and possibly with a whole suite of agents acting simultaneously) (67). None of these requirements is likely to make this enterprise any more tractable than it is at the moment.

There are advantages, however, in taking such a basic approach to understanding chemical defense allocation. First of all, expressing costs and benefits in terms of fitness couches the discussion in an evolutionary framework, a framework that is missing from many current discussions of plant–herbivore interactions. No discussion of adaptation can be purely ecological, since the process of adaptation is an evolutionary one; in all of the discussions of life history syndromes associated with defensive strategies (39–42, 59), virtually no evidence exists that any of the traits characterizing these syndromes are genetically based or, equally important, genetically correlated and likely to evolve in concert. Current patterns of allocation observed today are the result of an evolutionary process and are likely to change in the future as a result of evolutionary processes; it is difficult to appreciate ecological patterns without at least a rudimentary understanding of their evolutionary underpinnings and understanding the evolutionary process necessitates identifying selective agents and quantifying the selective forces they exert.

Restoring evolution to a place of prominence in future discussions of chemical defense means greatly increasing attention to the genetics of chemical defense production and allocation. Research pursuits in the study of chemical defense that should receive a renewed interest in this context include (*i*) investigation of multiple classes of secondary metabolites within a species, (*ii*) establishment of the genetic basis for chemical variability within a species, (*iii*) testing for toxicological interactions and genetic correlations, (*iv*) determination of genetic correlations between chemistry and life history traits, (*v*) precise estimation of consumer effects on fitness and influence of chemical variation on those effects, and (*vi*) elucidation of modes of action and mechanisms of genetically based counteradaptation in consumers. All of these pursuits require a sophisticated understanding of the biochemistry of a particular system, such that individual metabolites, not functional categories of secondary compounds, are identified, quantified, and monitored. The fact that secondary compounds may be involved in functions other than defense against consumers cannot be overlooked, particularly if expected genetic correlations fail to materialize.

An additional advantage of examining chemical defense allocation in the simplified context of benefit/cost analysis in particular systems is that such an approach is essentially unbiased relative to taxon. Theories derived from plant–herbivore interactions place an undue emphasis on the ability of plants to synthesize their primary metabolites from inorganic elements, but that ability may not necessarily be reflective, or predictive, of abilities to synthesize secondary metabolites. There is undeniably some degree of linkage between primary and secondary metabolism—at the very least, dead plants do not manufacture secondary metabolites—but there is little evidence to support the linear relationship that is assumed to prevail between them. If secondary metabolism cannot be totally divorced from primary metabolism in studying the ecology and evolution of chemical defense, irrespective of whether it takes place in plants, animals, or any other organisms, it may be productive for a while at least to arrange for a trial separation and see whether paradigms change.

SPECIAL CASE: HUMAN CHEMICAL DEFENSES

One compelling reason for examining patterns of chemical defense across taxa, and across trophic tiers, in order to distill out basic elements is that such an approach is essential in understanding chemical defense allocation in a seriously understudied species—*Homo sapiens*. By all rights and purposes, human beings should not utilize chemical defenses; as top carnivores in many food webs, they are rarely if ever consumed by other organisms [but see "Little Red Riding Hood," in Lang (68)]. Notwithstanding, humans have used every variant on chemical defense manifested by other organisms (Table 3). Like insects (69), parasitic plants (49–51), some birds (70), marine invertebrates (26, 67) and vertebrates (71), humans have coopted plant toxins to protect themselves against their consumers; the use of botanical preparations to kill insects, parasitic and otherwise, antedates written history (72). Many of the chemicals in use today as biocides, including antibiotics, are derived from other organisms. Natural products provide ≈25% of all drugs in use today (73) and botanical insecticides (or their chemically optimized derivatives) are still widely used as pest control agents (74). More recently, like many species of plants (4), insects (3), marine invertebrates (26), and vertebrates (27), humans have taken to synthesizing their own defensive compounds from inorganic materials, primary metabolites or from small precursor molecules, although this synthesis takes place in a laboratory or factory rather than in a cell or gland. In doing so, humans face many of the same challenges faced by other organisms that synthesize chemicals (Table 3)—in addition to the actual synthesis, the storage, transport, and

TABLE 3 Human chemical defense characteristics

Parallels between humans and other organisms	
Humans face	*Other organisms face*
Material and energy construction costs	Material and energy construction costs
Shipping, storage, and handling costs	Solubility, transportability, autotoxicity costs
Delivery costs, environmental impact, and autotoxicity considerations	Problems associated with delivery—toxicity to mutalists, increased visibility to specialists
"Opportunity costs"—alternative practices	"Opportunity costs"—alternative functions
Customer satisfaction associated with efficacy and versatility	Selection for mode of action, degree of efficacy, and range of organisms affected
Loss of sales due to resistance acquisition	Loss of efficacy due to consumer adaptation
Differences between humans and other organisms	
In humans, chemicals are used	*In other organisms, chemicals are used*
Prophylactically	Inducibly (elicited by damaging agent)
To kill	To minimize impact of consumer
In a broadcast fashion	In a tissue- or otherwise site-specific fashion
As single purified toxins	As variable mixtures

avoidance of autotoxicity all exact an economic cost, and the product is "selected" by consumers based on its efficacy and its range of uses relative to other products.

Although there are similarities between humans and other organisms in the acquisition of chemical defenses, there are striking differences in the deployment of these defenses (Table 3). Whereas most organisms use chemical defenses to minimize their own risk of being consumed, humans use chemicals in an offensive fashion, with the express purpose of killing off not only potential consumer organisms but also potential competitors for food or shelter. Throughout history, humans have even used chemicals to kill off conspecific competitors, a use of chemicals that is certainly unusual in the natural world (75). But humans have for the most part adopted the chemicals used by other organisms as defenses without at the same time investigating the manner in which these chemicals are deployed.

In general, chemical defenses in other organisms reflect the probability of attack and relative risk of damage (42); humans have in recent years developed a tendency to use chemical defenses prophylactically, even when potential enemies are absent. Humans tend to select those chemical agents that kill, rather than repel or misdirect, a large proportion of the target population; from a plant's perspective, the ultimate goal of chemical defense is to avoid being eaten, a goal that is as achievable if the plant is never consumed in the first place as it is if the consumer dies after ingesting a mouthful of toxin-laced tissue. Most organisms manufacture complex mixtures of chemicals for defense, some of which may actually be inactive as pure compounds (58); humans tend to prefer highly active individual components. It is perhaps the ecologically inappropriate deployment of these natural products (and their synthetic derivatives) by humans that has led to the widespread acquisition of resistance in all manner of target species and the concomitant loss of efficacy of these chemicals (76).

Efforts in recent years to identify natural sources of insecticidal and other pesticidal materials have increased, at least in part due to continuing problems with nontarget effects of synthetics and the widespread appearance of resistance to these compounds. Natural products are thought to offer greater biodegradability and possibly greater specificity than synthetic organic alternatives (77). However, the search for new biocidal agents is being conducted essentially as it has been done for the past century, with mass screening, isolation of active components, and development of syntheses for mass production of the active components (77, 78). Little or no effort is being expended by those interested in developing new chemical control agents in elucidating the manner in which plants or other source organisms manufacture, store, activate, transport, or allocate these chemicals. By understanding the rules according to which organisms defend themselves chemically, there is perhaps as much to be gained, in terms of developing novel and environmentally stable approaches to chemical control of insects and other pests, as there is by isolating and identifying the chemical defenses themselves.

SUMMARY

Defensive chemicals used by organisms for protection against potential consumers are generally products of secondary metabolism. Such chemicals are characteristic of free-living organisms with a limited range of movement or limited control over their movements. Despite the fact that chemical defense is widespread among animals as well as plants, the vast majority of theories advanced to account for patterns of allocation of energy and materials to defensive chemistry derive exclusively from studies of plant–herbivore interactions. Many such theories place an undue emphasis on primary physiological processes that are unique to plants (e.g., photosynthesis), rendering such theories limited in their utility or predictive power. The general failure of any single all-encompassing theory to gain acceptance to date may indicate that such a theory might not be a biologically realistic expectation. In lieu of refining theory, focusing attention on the genetic and biochemical mechanisms that underlie chemical defense allocation is likely to provide greater insights into understanding patterns across taxa. In particular, generalizations derived from understanding such mechanisms in natural systems have immediate applications in altering patterns of human use of natural and synthetic chemicals for pest control.

I thank Thomas Eisner and Jerrold Meinwald for asking me to think in broad terms about chemical defenses and for serving as an inspiration to me for the past two decades, and I thank Arthur Zangerl and James Nitao for their comments, insights, and unique good cheer. This work was supported in part by National Science Foundation Grant DEB 91-19612.

REFERENCES

1. Czapek, F. (1921) *Biochemie der Pflanzen* (Fischer, Jena, Germany), 1 Aufl. 3 Bd.
2. Lehninger, A. L. (1970) *Biochemistry* (Worth, New York).
3. Blum, M. (1981) *Chemical Defenses of Arthropods* (Academic, New York).
4. Berenbaum, M. R. & Seigler, D. (1993) in *Insect Chemical Ecology*, eds. Roitberg, B. & Isman, M. (Chapman & Hall, New York), pp. 89–121.
5. Murray, R. D., Mendez, H. J. & Brown, S. A. (1982) *The Natural Coumarins: Occurrence, Chemistry and Biochemistry* (Wiley, Chichester, U.K.).
6. Simsová, J. & Blazek, Z. (1967) *Çesk. Farm.* **16,** 22–28.
7. Zangerl, A. R. & Berenbaum, M. R. (1990) *Ecology* **71,** 1933–1940.
8. Berenbaum, M. R. (1981) *Oecologia* **49,** 236–244.
9. Nitao, J. K. & Zangerl, A. R. (1987) *Ecology* **68,** 521–529.
10. Berenbaum, M. R. & Zangerl, A. R. (1986) *Phytochemistry* **25,** 659–661.
11. Zangerl, A. R., Nitao, J. K. & Berenbaum, M. R. (1991) *Evol. Ecol.* **5,** 136–145.
12. Zangerl, A. R., Berenbaum, M. R. & Levine, E. (1989) *J. Hered.* **80,** 404–407.
13. Luckner, M. (1980) *J. Nat. Prod.* **43,** 21–40.
14. Rodriguez, E. & Levin, D. (1976) *Rec. Adv. Phytochem.* **10,** 214–271.
15. Nordlund, D. (1981) in *Semiochemicals: Their Role in Pest Control*, eds. Nordlund, D., Jones, R. & Lewis, W. (Wiley, New York), pp. 13–30.
16. Stowe, M. K. (1988) in *Chemical Mediation of Coevolution*, ed. Spencer, K. (Academic, New York), pp. 513–580.
17. Rice, E. (1977) *Biochem. Syst. Ecol.* **5,** 201–206.
18. Nielson, W. A., ed. (1936) *Webster's New International Dictionary of the English Language* (Merriam, Springfield, MA).
19. Seigler, D. S. & Price, P. (1976) *Am. Nat.* **110,** 101–105.
20. Seigler, D. S. (1977) *Biochem. Syst. Ecol.* **5,** 195–199.
21. Berenbaum, M. R. (1987) *ACS Symp. Ser.* **339,** 206–216.
22. Bernays, E. A. & Graham, M. (1988) *Ecology* **69,** 886–892.
23. Jones, C. G. & Firn, R. D. (1991) *Philos. Trans. R. Soc. London B* **333,** 273–280.
24. Jermy, T. (1993) *Entomol. Exp. Appl.* **66,** 3–12.
25. Dumbacher, J. P., Beehler, B. M., Spande, T. F., Barraffo, H. M. & Daly, J. W. (1992) *Science* **258,** 799–801.
26. Pawlik, J. R. (1993) *Chem. Rev.* **93,** 1911–1922.
27. Andersen, K. K. & Bernstein, D. T. (1975) *J. Chem. Ecol.* **1,** 493–499.
28. Bakus, G. J., Targett, N. M. & Schulte, B. (1986) *J. Chem. Ecol.* **12,** 951–987.
29. Lindquist, N., Hay, M. E. & Fenical, W. (1992) *Ecol. Monogr.* **62,** 547–568.

30. Eisner, T., Conner, W. E., Hicks, K., Dodge, K. R., Rosenberg, H. I., Jones, T. H., Cohen, M. & Meinwald, J. (1972) *Science* **196,** 1347–1349.
31. Hickman, C. P., Roberts, L. S. & Hickman, F. M. (1986) *Biology of Animals* (Times Mirror/Mosby, St. Louis).
32. Vaughan, T. (1972) *Mammalogy* (Saunders, Philadelphia).
33. Austin, O. L. (1961) *Birds of the World* (Hamlyn, New York).
34. Habermehl, G. (1981) *Venomous Animals and Their Toxins* (Springer, New York).
35. Price, P. W. (1991) *Oikos* **62,** 244–251.
36. White, T. C. R. (1974) *Oecologia* **16,** 279–301.
37. Levin, D. (1976) *Biochem. Syst. Ecol.* **6,** 61–76.
38. Atsatt, P. & O'Dowd, D. (1976) *Science* **193,** 24–29.
39. Feeny, P. (1976) *Rec. Adv. Phytochem.* **10,** 1–40.
40. Rhoades, D. F. & Cates, R. G. (1976) *Rec. Adv. Phytochem.* **10,** 168–213.
41. Rhoades, D. F. (1979) in *Herbivores: Their Interaction with Secondary Plant Metabolites,* eds. Rosenthal, G. & Janzen, D. (Academic, New York), pp. 3–54.
42. McKey, D. (1979) in *Herbivores: Their Interaction with Secondary Plant Metabolites,* eds. Rosenthal, G. & Janzen, D. (Academic, New York), pp. 55–133.
43. Bryant, J. P., Chapin, F. S. & Klein, D. R. (1983) *Oikos* **40,** 357–368.
44. Coley, P. D., Bryant, J. P. & Chapin, F. S. (1985) *Science* **230,** 895–899.
45. Bryant, J. P., Tuomi, J. & Niemelaa, P. (1988) in *Chemical Mediation of Coevolution,* ed. Spencer, K. C. (Academic, San Diego), pp. 367–389.
46. Price, P. W. (1991) *Ann. Entomol. Soc. Am.* **84,** 465–473.
47. Tuomi, J., Niemelaa, P., Haukioja, E., Siren, S. & Neuvonen, N. (1987) *Oecologia* **61,** 208–210.
48. Portier, P. (1949) *La Biologie des Lépidoptères* (Lechevalier, Paris).
49. Stermitz, F. R. & Harris, G. H. (1987) *J. Chem. Ecol.* **13,** 1917–1926.
50. Wink, M. & Witte, L. (1993) *J. Chem. Ecol.* **19,** 441–448.
51. Cordero, C. M., Serraon, A. M. G. & Gonzalez, M. J. A. (1993) *J. Chem. Ecol.* **19,** 2389–2393.
52. Reichardt, P. B., Chapin, F. S., Bryant, J. P., Mattes, B. R. & Clausen, T. P. (1991) *Oecologia* **88,** 401–406.
53. Lutz, L. (1928) *Bull. Soc. Bot. Fr.* **75,** 9–18.
54. Robinson, T. (1974) *Science* **184,** 430–435.
55. Lois, R., Dietrich, A., Hahlbrock, K. & Schulz, W. (1989) *EMBO J.* **8,** 1641–1648.
56. Ryan, C. A., Bishop, P. D., Graham, J. S., Broadway, R. M. & Duffey, S. S. (1986) *J. Chem. Ecol.* **12,** 1025–1036.
57. Gershenzon, J. (1994) *CRC Insect-Plant Interactions 4* (CRC, Boca Raton, FL), pp. 105–173.
58. Berenbaum, M. (1985) *Rec. Adv. Phytochem.* **19,** 139–169.
59. Herms, D. A. & Mattson, W. J. (1992) *Q. Rev. Biol.* **67,** 283–335.
60. Stamp, N. E. (1992) *Bull. Ecol. Soc. Am.* **73,** 28–39.
61. Simms, E. L. (1992) in *Plant Resistance to Herbivores and Pathogens,* eds. Simms, E. L. & Fritz, R. L. (Univ. of Chicago Press, Chicago), pp. 392–425.
62. Ågren, J. & Schemske, D. W. (1993) *Am. Nat.* **141,** 338–350.
63. Fagerstrom, T. (1989) *Am. Nat.* **133,** 281–287.
64. Simms, E. L. & Rausher, M. D. (1989) *Evolution (Lawrence, Kans.)* **43,** 573–585.
65. Coley, P. D. (1983) *Oecologia* **70,** 238–241.
66. Berenbaum, M. R. (1986) in *Insect-Plant Interactions,* eds. Miller, T. A. & Miller, J. (Springer, New York), pp. 121–153.
67. Hay, M. E. & Steinberg, P. D. (1992) in *Herbivores: Their Interactions with*

Secondary Plant Metabolites, eds. Rosenthal, G. & Berenbaum, M. (Academic, New York), pp. 371–413.
68. Lang, A. (1967) *The Blue Fairy Book* (Dover, New York).
69. Berenbaum, M. R. (1993) *Food Insects Newsl.* **6,** 1, 6–9.
70. Clark, L. & Mason, J. R. (1988) *Oecologia* **77,** 174–180.
71. Yasumoto, T. & Murata, M. (1993) *Chem. Rev.* **93,** 1897–1909.
72. Smith, A. E. & Secoy, D. M. (1975) *J. Agric. Food Chem.* **23,** 1050–1056.
73. Farnsworth, N. R. & Soejarto, D. D. (1985) *Econ. Bot.* **39,** 231–240.
74. Arnason, J. T., Philogene, B. J. R. & Morand, P., eds. (1989) *ACS Symp. Ser.* **387**.
75. Haber, L. (1986) *The Poisonous Cloud* (Oxford Univ. Press, Oxford).
76. Berenbaum, M. R. (1991) *Oxf. Rev. Evol. Biol.* **7,** 285–307.
77. Jacobsen, M. (1991) *ACS Symp. Ser.* **387,** 1–10.
78. Ku, H. S. (1987) *ACS Symp. Ser.* **330,** 449–454.
79. Daly, J. W. (1995) *Proc. Natl. Acad. Sci. USA* **92,** 9–13.

The Chemistry of Poisons in Amphibian Skin

JOHN W. DALY

Poisonous substances occur throughout nature and are particularly well known from plants, where they presumably serve in chemical defense against herbivores. Poisons can also serve as venoms, which are introduced into victims by coelenterates; molluscs; various arthropods, including insects, spiders, and scorpions; gila monsters; and snakes, by a bite or sting, or as toxins, such as those produced by bacteria, dinoflagellates, and other microorganisms. Examples of poisons of plant origin encompass a wide range of substances, including many alkaloids; a variety of terpenes and steroids, some of which occur as saponins; and unusual secondary metabolites such as the trichothecenes, pyrethroids, and dianthrones (1, 2). Another wide range of presumably defensive substances occur in marine invertebrates, including steroid and terpenoid sapogenins, tetrodotoxins, a variety of polyether toxins, and alkaloids (3, 4). Poisons also occur in terrestrial invertebrates and vertebrates, where they serve as chemical defenses by insects and other arthropods (5, 6), by fish (7), and by amphibians (8). Recently, a toxic alkaloid was characterized from the skin and feathers of a bird (9), where it confers some protection against predation by humans. Chemical defenses can be directed either against predators or against microorganisms. The present paper is concerned with the chemical nature, origin, and function of poisons present in amphibian skin. Many of the substances in amphibians

John Daly is chief, Laboratory of Bioorganic Chemistry, at the National Institutes of Health, Bethesda, Maryland.

might better be categorized as "noxious" rather than "poisonous," although at high enough dosages all of these compounds would be poisons.

Toads and salamanders have been considered noxious creatures for centuries and indeed the majority of amphibians have now been found to contain noxious and sometimes poisonous substances in their skin secretions (8). The type of biologically active substance found in amphibians appears to have phylogenetic significance. Thus, indole alkylamines are typically present in high levels in bufonid toads of the genus *Bufo*, phenolic amines in leptodactylid frogs, vasoactive peptides in a great variety in hylid frogs, particularly of genus *Phylomedusa* (10), and bufadienolides in parotoid glands and skins of bufonid toads of the genus *Bufo* as well as in skin of related bufonid genera *Atelopus* and probably *Dendrophryniscus* and *Melanophryniscus* (11). The water-soluble alkaloid tetrodotoxin occurs in newts of the family Salamandridae, toads of the brachycephalid genus *Brachycephalus* and the bufonid genus *Atelopus*, and now in one frog species of the dendrobatid genus *Colostethus* (12). Lipophilic alkaloids have been found only in salamanders of the salamandrid genus *Salamandra*; in frogs of the dendrobatid genera *Phyllobates*, *Dendrobates*, *Epipedobates*, and *Minyobates*, the mantellid genus *Mantella* and the myobatrachid genus *Pseudophryne*; and in toads of the bufonid genus *Melanophryniscus*. More than 70 other genera from 11 amphibian families do not have skin alkaloids. The distribution of various lipophilic alkaloids in amphibians is given in Table 1 and structures are shown in Figure 1.

The origin and function of poisons and noxious substances found in amphibians are only partially known. The high levels of amines, including such well-known biogenic amines as serotonin, histamine, and tyramine and derivatives thereof, found in skin of various toads and frogs (8), undoubtedly are synthesized by the amphibian itself. They are stored in granular skin glands for secretion upon attack by a predator, whereupon their well-known irritant properties on buccal tissue would serve well in chemical defense. The high levels of vasoactive peptides, such as bradykinin, sauvagine, physaelaemin, caerulein, bombesin, dermorphins, etc., presumably also serve in defense against predators, although many, including the magainins, have high activity as antimicrobials (13) and thus might also serve as a chemical defense against microorganisms. Skin secretions from one hylid frog are used in "hunting magic" folk rituals by Amazonian Indians; such secretions contain many vasoactive peptides (10) and a peptide, adenoregulin, that can affect central adenosine receptors (14). The peptides of frog skin are synthesized by the amphibian and indeed additional peptides are being deduced based on cDNAs for their precursors (15). The various hemolytic proteins of certain amphib-

TABLE 1 Occurrence of lipid-soluble alkaloids in amphibians

Family and genus	SAM	BTX	HTX	PTX-A class			DHQ	Izidine alkaloid				Epi	Pseudophry
				PTX	aPTX	hPTX		3,5-P	3,5-I	5,8-I	1,4-Q		
Salamandridae													
Salamandra	+	–	–	–	–	–	–	–	–	–	–	–	–
Dendrobatidae													
Phyllobates	–	+	+	+	–	–	+	–	+	–	–	–	–
Dendrobates	–	–	+	+	+	+	+	+	+	+	+	–	–
Epipedobates	–	–	+	+	+	–	+	–	–	+	+	+	–
Minyobates	–	–	–	+	+	–	+	–	–	+	+	–	–
Mantellidae													
Mantella	–	–	–	+	+	+	+	+	+	+	+	–	–
Myobatrachidae													
Pseudophryne	–	–	–	+	+	–	–	–	–	–	–	–	+
Bufonidae													
Melanophryniscus	–	–	–	+	+	+	+	+	+	+	+	–	–

SAM, samandarines; BTX, batrachotoxins; HTX, histrionicotoxins; PTX, pumiliotoxins; aPTX, allopumiliotoxins; hPTX, homopumiliotoxins; DHQ, 2,5-disubstituted decahydroquinolines; 3,5-P, 3,5-disubstituted pyrrolizidines; 3,5-I and 5,8-I, disubstituted indolizidines; 1,4-Q, 1,4-disubstituted quinolizidines; Epi, epibatidine; Pseudophry, pseudophrynamines. With the exception of 3,5-P and 3,5-I, these alkaloids are not known to occur in arthropods (see text). Histrionicotoxins may occur in *Minyobates* and *Mantella*, but the evidence is not conclusive.

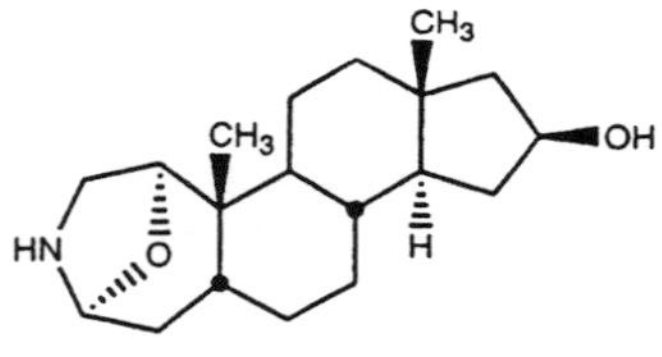

Samandarine*

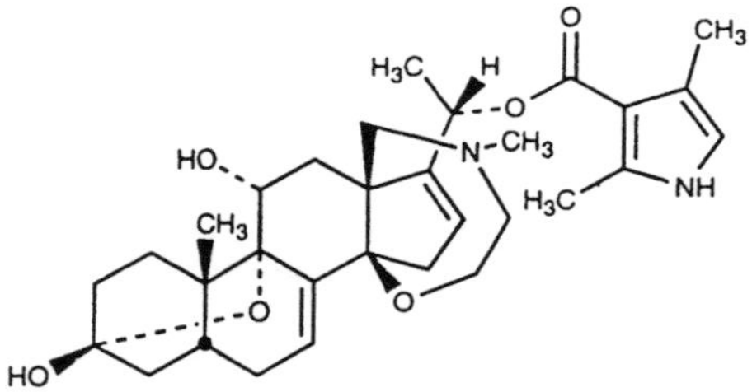

Batrachotoxin*

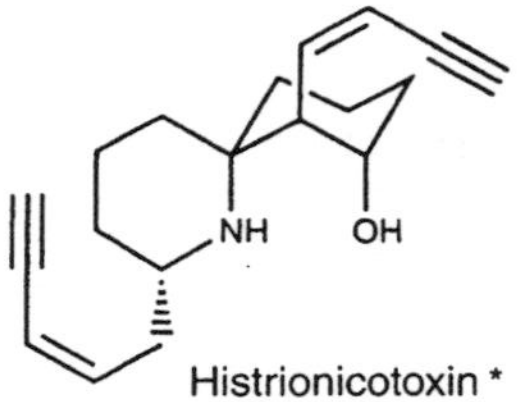

Histrionicotoxin*

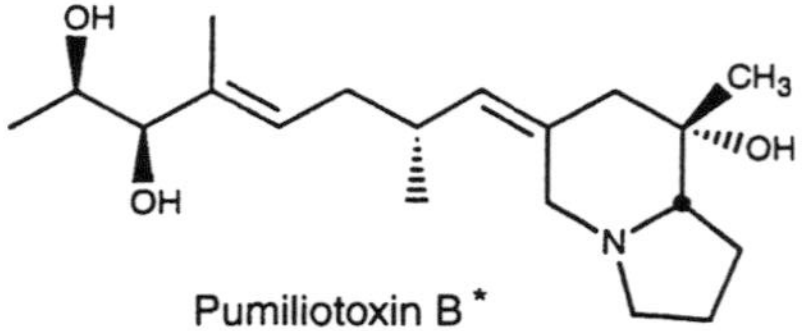

Pumiliotoxin B*

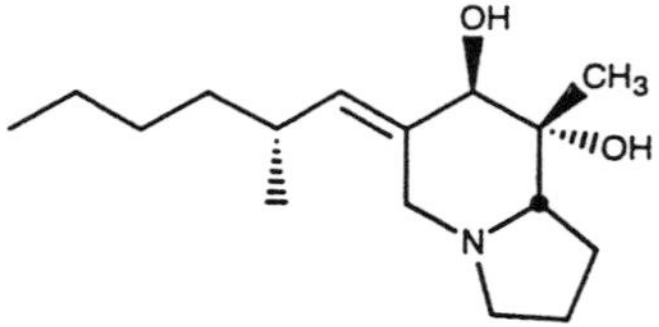

Allopumiliotoxin **267A***

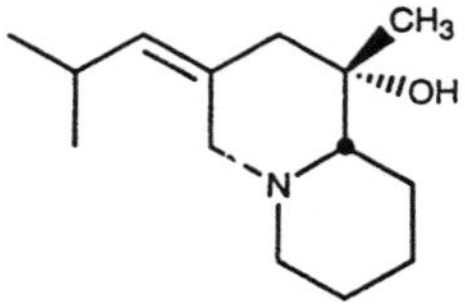

Homopumiliotoxin **223G***

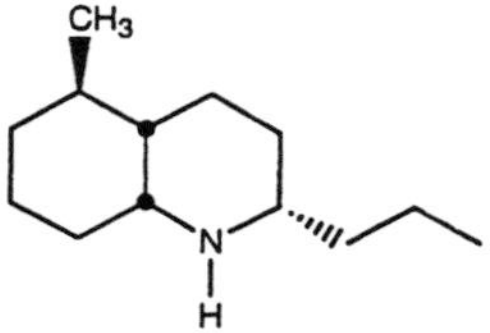

Decahydroquinoline **195A***

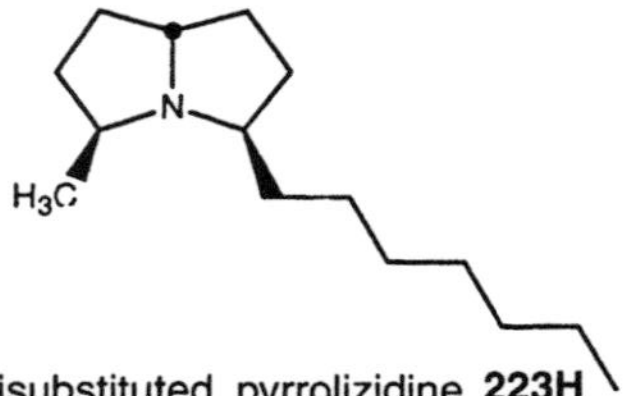

3,5-Disubstituted pyrrolizidine **223H**

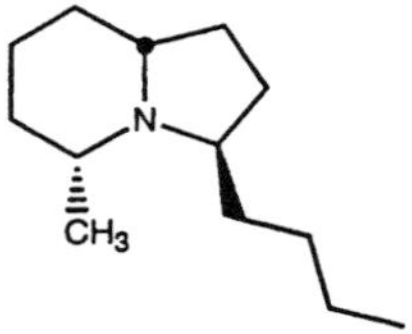

3,5-Disubstituted indolizidine **195B**

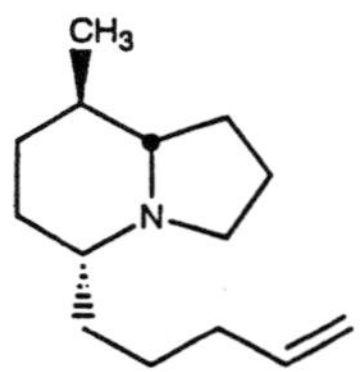

5,8-Disubstituted indolizidine **207A***

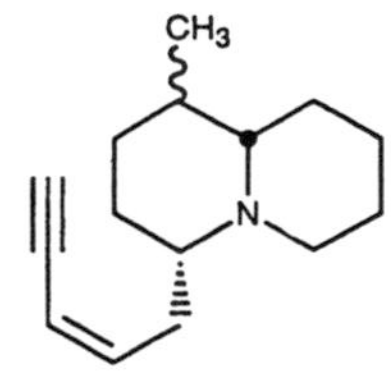

1,4-Disubstituted quinolizidine **217A***

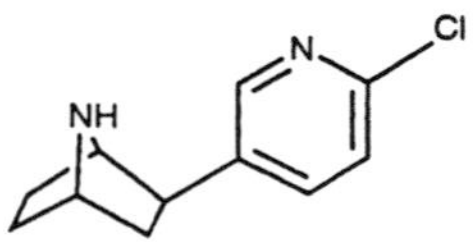

Epibatidine*

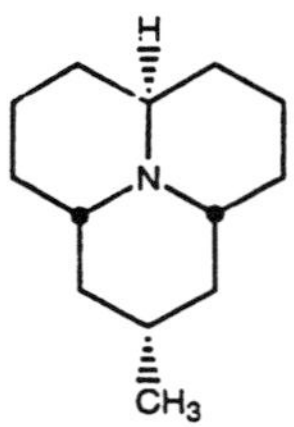

Precoccinelline

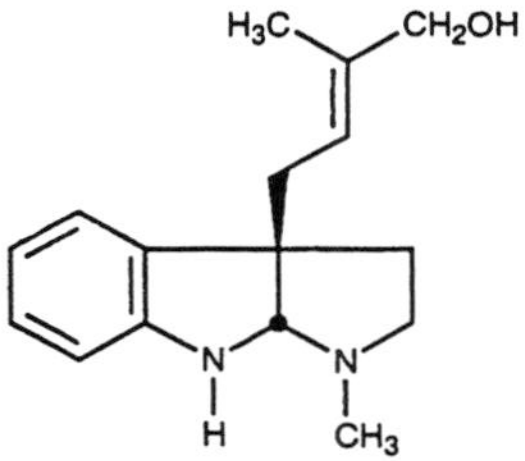

Pseudophrynaminol*

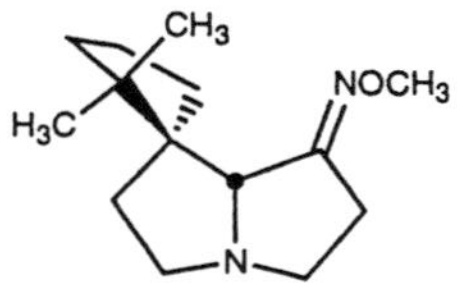

Pyrrolizidine oxime **236**

FIGURE 1 Structures of lipophilic amphibian alkaloids. Alkaloids indicated by asterisks represent structural classes that have not been detected in nature except in amphibians and, in the case of batrachotoxins, in one species of bird (9).

ians are certainly of endogenous origin. The steroidal bufadienolides appear to be synthesized from cholesterol by the bufonid toads (16). It has been suggested that structurally similar and toxic lucibufagins of fireflies might also be produced by the insect from dietary cholesterol (17). However, toxic cardenolides in monarch butterflies appear to be sequestered from milkweed plants by the larvae (18). The chemical defensive attributes of the highly toxic bufadienolides are due to effects on membrane Na^+/K^+-ATPase.

The origin of tetrodotoxins in amphibians and higher organisms remains enigmatic. Thus, puffer fish raised in hatcheries do not contain tetrodotoxin (19), and likely biosynthetic precursors are not incorporated into tetrodotoxin with newts (20). Feeding nontoxic puffer fish with tetrodotoxin does not result in sequestration, but feeding toxic ovaries from wild puffer fish does (19). A bacterial origin for tetrodotoxin has been suggested, but such a source fails to explain the fact that one Central American species of toad of the genus *Atelopus* contains mainly tetrodotoxin; another Central American species contains mainly chiriquitoxin, which is a unique but structurally similar toxin; and yet another contains mainly zetekitoxin, which is another unique, probably structurally related toxin (see ref. 12). Chiriquitoxin, while related in structure to tetrodotoxin, differs in the carbon skeleton (21). The chemical defensive attributes of tetrodotoxin are due to blockade of voltage-dependent sodium channels and hence cessation of neuronal and muscle activity.

The origin of the lipophilic alkaloids in dendrobatid frogs, engendered by the observation that the frogs, which are used by Colombian Indians to poison blow darts, when raised in captivity contain none of the toxic batrachotoxins present in wild-caught frogs (22), remains to be investigated. In contrast, the toxic samandarines from fire salamanders are present in the skin glands of the salamander through many generations of nurture in captivity (G. Habermehl, personal communication). The various lipophilic alkaloids of amphibians all have marked activity on ion channels and hence through such effects would serve effectively as chemical defenses, even though some have relatively low toxicity.

The batrachotoxins were the first class of unique alkaloids to be characterized from skin extracts of frogs of the family Dendrobatidae (see ref. 23 for a review of amphibian alkaloids). Batrachotoxin was detected in only five species of dendrobatid frogs and these frogs were then classified as the monophyletic genus *Phyllobates*, based in part on the presence of batrachotoxins (24). However, levels of batrachotoxins differ considerably, with the Colombian *Phyllobates terribilis* containing nearly 1 mg of batrachotoxins per frog, while the somewhat smaller *Phyllobates bicolor* and *Phyllobates aurotaenia*, also from the rain forests of the Pacific versant in Colombia, contain 10-fold lower skin levels (8). The two

Phyllobates species from Panama and Costa Rica contain either only trace amounts of batrachotoxin or for certain populations of *Phyllobates lugubris* no detectable amounts. Batrachotoxins are unique steroidal alkaloids, which were unknown elsewhere in nature until the recent discovery of homobatrachotoxin at low levels in skin and feathers of a Papua New Guinean bird of the genus *Pitohui* (9). In the dendrobatid frogs, three major alkaloids are present—namely, batrachotoxin, homobatrachotoxin, and a much less toxic possible precursor, batrachotoxinin A. The latter, when fed to nontoxic captive-raised *P. bicolor* using dusted fruit flies, is accumulated into skin glands but is not converted to the more toxic esters batrachotoxin and homobatrachotoxin (25). Dendrobatid frogs of another genus would not eat the batrachotoxinin-dusted fruit flies. Batrachotoxins depolarize nerve and muscle by specific opening of sodium channels; the sodium channels of the *Phyllobates* species are insensitive to the action of batrachotoxin (22).

Further examination of extracts of dendrobatid frogs over nearly 3 decades led to the characterization of nearly 300 alkaloids, representing some 18 structural classes (see ref. 23). Several classes remain unknown in nature except in frog skin (see Table 1), and their origin remains obscure in view of the relatively recent finding that frogs of the dendrobatid genera *Dendrobates* and *Epipedobates*, like *Phyllobates*, do not have skin alkaloids when raised in captivity (26). The distribution of the various alkaloids of amphibians is pertinent to any speculation as to their origin (see Table 1).

The so-called pumiliotoxin A class of "dendrobatid alkaloids" is as yet known only in nature from frog/toad skin. The class consists of alkaloids with either an indolizidine (pumiliotoxins and allopumiliotoxins) or a quinolizidine (homopumiliotoxins) ring, in each case with a variable alkylidene side chain. The pumiliotoxin A class occurs in skin of all of the amphibian genera that contain lipophilic alkaloids with the exception of the fire salamanders, which contain only samandarines. In spite of a wide distribution in the alkaloid-containing frogs, there are species and/or populations of frogs that have no pumiliotoxin A class alkaloids or only trace amounts. Members of pumiliotoxin A class are active toxins with effects on sodium and perhaps calcium channels and, thus, would serve well in defense against predators.

Histrionicotoxins represent another major class of dendrobatid alkaloids. They contain a unique spiropiperidine ring system and side chains with acetylenic, olefinic, and allenic groups. Histrionicotoxins remain known in nature only from dendrobatid frogs of the general *Phyllobates, Dendrobates*, and *Epipedobates*. They are probably absent in the tiny dendrobatid frogs of the genus *Minyobates*. Histrionicotoxins were detected in a single Madagascan frog of the mantellid genus *Mantella* (27) obtained through the pet trade but have not been detected in any extracts of several *Mantella* species collected in Madagascar (28). Histrionicotoxins do not occur in all species of the above dendrobatid frog genera or in all populations of a single species (8). Their occurrence within populations of a species on a single small island can vary from high levels to none.

The decahydroquinolines are the third major class of dendrobatid alkaloids still known only from frog/toad skin. Decahydroquinolines occur in skin of all the frog/toad genera that have lipophilic alkaloids with the sole exception of the Australian myobatrachid frogs of the genus *Pseudophryne* that contain only (allo)pumiliotoxins and a series of indole alkaloids unique in nature to this genus of frogs—namely, the pseudophrynamines (29).

A series of simple bicyclic alkaloids could be considered to make up a major "izidine" class of alkaloids in the dendrobatid and other frogs. These include the 3,5-disubstituted pyrrolizidines, the 3,5-disubstituted and 5,8-disubstituted indolizidines, and the 1,4-disubstituted quinolizidines. The 3,5-disubstituted pyrrolizidines and the 3,5-disubstituted indolizidines are not unique to frogs, having been reported from ants (see ref. 6). Ants thus represent a potential dietary source for such alkaloids in dendrobatid and other frogs. Indeed, feeding experiments with ants of the genus *Monomorium* that contain a 3,5-disubstituted indolizidine and a 2,5-disubstituted pyrrolidine resulted in a remarkable selective accumulation, into the skin of the dendrobatid frog *Dendrobates auratus*, of the indolizidine but not of the pyrrolidine (25). It should be noted that some dendrobatid frogs do contain significant levels in skin of such 2,5-disubstituted pyrrolidines and of 2,6-disubstituted piperidines, neither of which appears to be sequestered into skin, at least by *D. auratus*. The 5,8-disubstituted indolizidines and 1,4-disubstituted quinolizidines remain as yet unknown in nature except from frog/toad skin (Table 1).

There are also a number of alkaloids characterized from skin extracts of dendrobatid frogs that have a rather limited distribution within the many species that have been examined and are as yet known only from frog skin. The tricyclic gephyrotoxins occur along with the more widely distributed histrionicotoxins in only a few species and populations of dendrobatid frogs (8). The tricyclic cyclopenta[*b*]quinolizidines occur in only one species, a tiny Colombian frog *Minyobates bombetes* (30). The potent nicotinic analgesic epibatidine occurs only in four dendrobatid species of the genus *Epipedobates* found in Ecuador (31).

Two classes of dendrobatid alkaloids have potential dietary sources. The first are the pyrrolizidine oximes (32), whose carbon skeleton is identical to that of nitropolyzonamine, an alkaloid from a small millipede (33). Indeed, raising the dendrobatid frog *D. auratus* in Panama on leaf-litter arthropods, gathered weekly, resulted in skin levels of the pyrrolizidine oxime **236** even higher than levels in wild-caught frogs from the leaf-litter site (34). The second are the tricyclic coccinelline alkaloids that have been found in several frogs/toads. The coccinellines occur as defensive substances in a variety of small beetles (see ref. 6). Thus, beetles represent a possible dietary source for coccinelline-class alkaloids in frog/toad skin. Indeed, the beetle alkaloid precoccinelline is a significant alkaloid in the skin of *D. auratus* raised in Panama on leaf-litter arthropods (34). The other alkaloids that were found in skin of *D. auratus* raised on leaf-litter arthropods are three other tricyclic alkaloids, perhaps of the coccinelline class but of unknown structure, two 1,4-disubstituted quinolizidines, a gephyrotoxin, a decahydroquinoline, and several histrionicotoxins. With the exception of the pyrrolizidine oxime **236**, skin levels of the various alkaloids in the captive-raised frogs were low compared to levels of alkaloids in wild-caught frogs from the leaf-litter collection site or from the parental stock of *D. auratus* on a nearby island (34). Individual variation in wild-caught frogs appears significant, which complicates the comparisons. However, the lack of any pumiliotoxins and the relatively low levels or absence of decahydroquinolines and histrionicotoxins in the captive-raised frogs suggests that dietary sources for these alkaloids have been missed in the paradigm using large funnels to collect the arthropods from the leaf litter.

In summary, poisons used in chemical defense are widespread in nature. In amphibians, the defensive substances seem to be elaborated by the amphibian in the case of amines, peptides, proteins, bufadienolides, and the salamander alkaloids of the samandarine class. For the tetrodotoxin class of water-soluble alkaloids, the origin is unclear, but symbiotic bacteria have been suggested for marine organisms (4). For the so-called dendrobatid alkaloids, a dietary source now appears a likely explanation for the lack of skin alkaloids in dendrobatid frogs raised in captivity. Certainly, dendrobatid frogs of the dendrobatid genera *Phyllobates*, *Dendrobates*, and *Epipedobates*, which in the wild contain skin alkaloids, have highly efficient systems for accumulating selectively into skin a variety of dietary alkaloids (25, 34). A biological system for sequestration of alkaloids for chemical defense finds precedence in the transfer of pyrrolizidine alkaloids from plants via aphids to ladybug beetles (35). Accumulation of cantharidins in muscle of ranid frogs after feeding on beetles has been documented (36). Frogs of the dendrobatid genus *Colostethus*, which in the wild do not contain skin alkaloids, do not accumulate dietary alkaloids (25).

The proposal that all alkaloids found in skin glands of dendrobatid frogs and used in chemical defense against predators have a dietary origin leads to many questions. First, the profile of alkaloids has been found in many instances to be characteristic of a species or a population.

Thus, either the systems responsible for sequestration of alkaloids differ in selectivity among different species and/or populations of dendrobatid frogs or the small arthropod fauna presenting itself and used as a diet by different species and/or populations varies even within a small island. The latter appears more likely. It was noted that the dendrobatid frogs raised on leaf litter in Panama shared more alkaloids with a population of *D. auratus* from the leaf-litter site than they did with the parental population from a nearby island (34). The second major question concerns what small insects or other arthropods contain such toxic and/or unpalatable alkaloids as the batrachotoxins, the pumiliotoxins, and the histrionicotoxins, the decahydroquinolines, the 5,8-disubstituted indolizidines, the 1,4-disubstituted quinolizidines, and epibatidine. It is remarkable that such small, presumably distasteful arthropods have escaped the attention of researchers. Whether frogs intent on sequestering defensive alkaloids seek out such prey is unknown. With regard to the frogs/toads from the Madagascan family Mantellidae, the Australian family Myobatrachidae and the South American genus *Melanophryniscus* of the family Bufonidae, which also contain many of the dendrobatid alkaloids, it is unknown whether sequestering systems are present or even whether captive-raised frogs will lack skin alkaloids. If such systems are present, then it is remarkable from an evolutionary standpoint that such unrelated lineages of toads/frogs have independently developed systems for sequestering alkaloids into skin glands from a diet of small, presumably noxious insects for use by the toad/frog in chemical defense.

SUMMARY

Poisons are common in nature, where they often serve the organism in chemical defense. Such poisons either are produced *de novo* or are sequestered from dietary sources or symbiotic organisms. Among vertebrates, amphibians are notable for the wide range of noxious agents that are contained in granular skin glands. These compounds include amines, peptides, proteins, steroids, and both water-soluble and lipid-soluble alkaloids. With the exception of the alkaloids, most seem to be produced *de novo* by the amphibian. The skin of amphibians contains many structural classes of alkaloids previously unknown in nature. These include the batrachotoxins, which have recently been discovered to also occur in skin and feathers of a bird, the histrionicotoxins, the gephyrotoxins, the decahydroquinolines, the pumiliotoxins and homopumiliotoxins, epibatidine, and the samandarines. Some amphibian skin alkaloids are clearly sequestered from the diet, which consists mainly of small arthropods. These include pyrrolizidine and indolizidine alkaloids from ants, tricyclic coccinellines from beetles, and pyrrolizidine oximes, presumably from millipedes. The sources of other alkaloids in amphibian skin, including the batrachotoxins, the decahydroquinolines, the histrionicotoxins, the pumiliotoxins, and epibatidine, are unknown. While it is possible that these are produced *de novo* or by symbiotic microorganisms, it appears more likely that they are sequestered by the amphibians from as yet unknown dietary sources.

REFERENCES

1. Whittaker, R. H. & Feeny, P. P. (1971) *Science* **171,** 757–770.
2. Balandrin, M. F., Klocke, J. A., Wurtele, E. S. & Bollinger, W. H. (1985) *Science* **228,** 1154–1160.
3. Scheuer, P. J. (1990) *Science* **248,** 173–177.
4. Yasumoto, T. & Murata, M. (1993) *Chem. Rev.* **93,** 1897–1909.
5. Eisner, T. & Meinwald, J. (1966) *Science* **153,** 1341–1350.
6. Jones, T. H. & Blum, M. S. (1983) *Alkaloids: Chemical and Biological Perspectives* (Wiley, New York), Vol. 1, pp. 33–84.
7. Tachibana, K. (1988) *Bioorg. Mar. Chem.* **2,** 124–145.
8. Daly, J. W., Myers, C. W. & Whittaker, N. (1987) *Toxicon* **25,** 1023–1095.
9. Dumbacher, J. P., Beehler, B. M., Spande, T. F., Garraffo, H. M. & Daly, J. W. (1992) *Science* **258,** 799–801.
10. Erspamer, V., Melchiorri, P., Erspamer, G. F., Montecucchi, P. C. & De Castiglione, R. (1985) *Peptides* **6,** 7–12.
11. Flier, J., Edwards, M. W., Daly, J. W. & Myers, C. W. (1980) *Science* **208,** 503–505.
12. Daly, J. W., Gusovsky, F., Myers, C. W., Yotsu-Yamashita, M. & Yasumoto, T. (1994) *Toxicon* **32,** 279–285.
13. Bevins, C. L. & Zasloff, M. (1990) *Annu. Rev. Biochem.* **59,** 395–414.
14. Daly, J. W., Caceres, J., Moni, R. W., Gusovsky, F., Moos, M., Jr., Seamon, K. B., Milton, K. & Myers, C. W. (1992) *Proc. Natl. Acad. Sci. USA* **89,** 10960–10963.
15. Richter, K., Egger, R., Negri, L., Corsi, R., Severini, C. & Kreil, G. (1990) *Proc. Natl. Acad. Sci. USA* **87,** 4836–4839.
16. Porto, A. M. & Gros, E. (1971) *Experientia* **27,** 506.
17. Eisner, T., Wiemer, D. F., Haynes, L. W. & Meinwald, J. (1978) *Proc. Natl. Acad. Sci. USA* **75,** 905–908.
18. Brower, L. P., van Zandt Brower, J. & Corvino, J. M. (1967) *Proc. Natl. Acad. Sci. USA* **57,** 893–898.
19. Matsui, T., Hamada, S. & Konosu, S. (1981) *Bull. Jpn. Soc. Sci. Fish.* **47,** 535–537.
20. Shimizu, Y. & Kobayashi, M. (1983) *Chem. Pharm. Bull.* **31,** 3625–3631.
21. Yotsu, M., Yasumoto, T., Kim, Y. H., Naoki, H. & Kao, C. Y. (1990) *Tetrahedron Lett.* **31,** 3187–3190.
22. Daly, J. W., Myers, C. W., Warnick, J. E. & Albuquerque, E. X. (1980) *Science* **208,** 1383–1385.
23. Daly, J. W., Garraffo, H. M. & Spande, T. F. (1993) in *The Alkaloids*, ed. Cordell, G. A. (Academic, San Diego), Vol. 43, Chap. 3, pp. 185–288.
24. Myers, C. W., Daly, J. W. & Malkin, B. (1978) *Bull. Am. Mus. Nat. Hist.* **161,** 307–365.
25. Daly, J. W., Secunda, S. I., Garraffo, H. M., Spande, T. F., Wisnieski, A. & Cover, J. F., Jr. (1994) *Toxicon* **32,** 657–663.

26. Daly, J. W., Secunda, S. I., Garraffo, H. M., Spande, T. F., Wisnieski, A., Nishihira, C. & Cover, J. F., Jr. (1992) *Toxicon* **30,** 887–898.
27. Daly, J. W., Highet, R. J. & Myers, C. W. (1984) *Toxicon* **22,** 905–919.
28. Garraffo, H. M., Caceres, J., Daly, J. W., Spande, T. F., Andriamaharavo, N. R. & Andriantsiferana, M. (1993) *J. Nat. Prod.* **56,** 1016–1038.
29. Daly, J. W., Garraffo, H. M., Pannell, L. K. & Spande, T. F. (1990) *J. Nat. Prod.* **53,** 407–421.
30. Spande, T. F., Garraffo, H. M., Yeh, H. J. C., Pu, Q.-L., Pannell, L. K. & Daly, J. W. (1992) *J. Nat. Prod.* **55,** 707–722.
31. Spande, T. F., Garraffo, H. M., Edwards, M. W., Yeh, H. J. C., Pannell, L. & Daly, J. W. (1993) *J. Am. Chem. Soc.* **114,** 3475–3478.
32. Tokuyama, T., Daly, J. W., Garraffo, H. M. & Spande, T. F. (1992) *Tetrahedron* **48,** 4247–4258.
33. Meinwald, J., Smolanoff, J., McPhail, A. T., Miller, R. W., Eisner, T. & Hicks, K. (1975) *Tetrahedron Lett.* 2367–2370.
34. Daly, J. W., Garraffo, H. M., Spande, T. F., Jaramillo, C. & Rand, A. S. (1994) *J. Chem. Ecol.* **4,** 943–955.
35. Witte, L., Ehmke, A. & Hartmann, T. (1990) *Naturwissenschaften* **77,** 540–543.
36. Eisner, T., Conner, J., Carrel, J. E., McCormick, J. P., Slagle, A. J., Gans, C. & O'Reilly, J. C. (1990) *Chemoecology* **1,** 57–62.

The Chemistry of Phyletic Dominance

JERROLD MEINWALD AND THOMAS EISNER

Whether we chose as our criterion Erwin's generous estimate of 30 million species of insects (1) or the somewhat more modest number favored by Wilson (2), it is clear that insects have achieved formidable diversity on Earth. On dry land they literally reign supreme. It has been estimated that there are some 200 million insects for each human alive (3). The eminence of the phylum Arthropoda among animals is very much a reflection of the success of the insects alone.

A number of factors have contributed to the achievement of dominance by insects. They were the first small animals to colonize the land with full success, thereby gaining an advantage over latecomers. Their exoskeleton shielded them from desiccation and set the stage for the evolution of limbs, tracheal tubes, and elaborate mouthparts, adaptations that were to enable insects to become agile, energetically efficient, and extraordinarily diverse in their feeding habits. Metamorphosis opened the option for insects to exploit different niches during their immature and adult stages. The evolution of wings facilitated their dispersal as adults, as well as their search for mates and oviposition sites. And there were subtle factors such as the acquisition of a penis (the aedeagus), which enabled insects to effect direct sperm transfer from male to female, without recourse to an outer aquatic environment for fertilization (3).

Jerrold Meinwald is Goldwin Smith Professor of Chemistry and Thomas Eisner is Schurman Professor of Biology and director of the Cornell Institute for Research in Chemical Ecology at Cornell University, Ithaca, New York.

A factor not often appreciated that also contributed to insect success is the extraordinary chemical versatility of these animals. Insects produce chemicals for the most diverse purposes—venoms to kill prey, repellents and irritants to fend off enemies, and pheromones for sexual and other forms of communication. The glands responsible for production of these substances are most often integumental, having been derived by specialization of localized regions of the epidermis. In insects, as in arthropods generally, the epidermis is fundamentally glandular, being responsible for secretion of the exoskeleton. In their acquisition of special glands, arthropods appear to have capitalized upon the ease with which epidermal cells can be reprogrammed evolutionarily for performance of novel secretory tasks. The glandular capacities of arthropods are known to anyone who has collected these animals in the wild. Arthropods commonly have distinct odors and are often the source of visible effluents that they emit when disturbed (Figure 1). Much has been learned in recent years about the function and chemistry of arthropod secretions. The insights gained have been fundamental to the emergence of the field of chemical ecology. Our purpose here is to focus on some aspects of the secretory chemistry of these animals, with emphasis on a few recent discoveries.

ARTHROPOD CHEMICAL DEFENSES

In our earliest collaborative publication, we described the dramatic chemical defense mechanism of the whip scorpion, *Mastigoproctus giganteus* (6). This ancient arachnid is able to spray a well-aimed stream of ≈85% acetic acid [CH_3CO_2H] containing 5% caprylic acid [$CH_3(CH_2)_6CO_2H$] at an assailant. The role of the caprylic acid proved to be especially interesting: it facilitates transport of the acetic acid through the waxed epicuticle of an enemy arthropod. It is likely that this simple strategy of using a lipophilic agent to enable a more potent compound to penetrate a predator's cuticle has helped this species to survive for as long as 400 million years. Many other independently evolved arthropod defensive secretions make use of the same strategy (9). The whip scorpion defense mechanism helped us to appreciate the fact that chemistry need not be complex to be effective. The virtue of simplicity is further illustrated by the defensive use of mandelonitrile and benzoyl cyanide, easily decomposed precursors of hydrogen cyanide (HCN), by certain millipedes and centipedes (10, 11).

While arthropod defensive secretions often rely for their effect on well-known aliphatic acids, aldehydes, phenols, and quinones, there are many cases in which compounds capable of whetting the appetite of any natural products chemist are utilized. For example, steroids play a

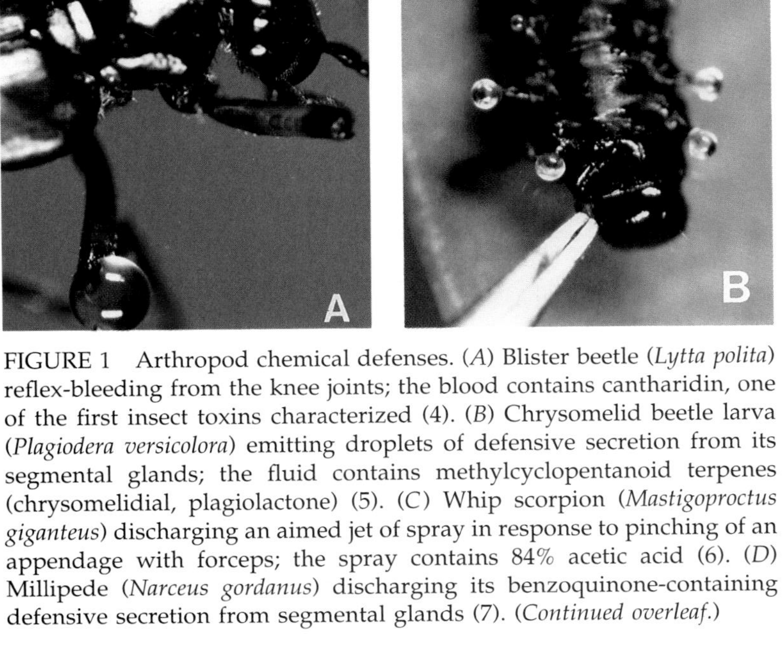

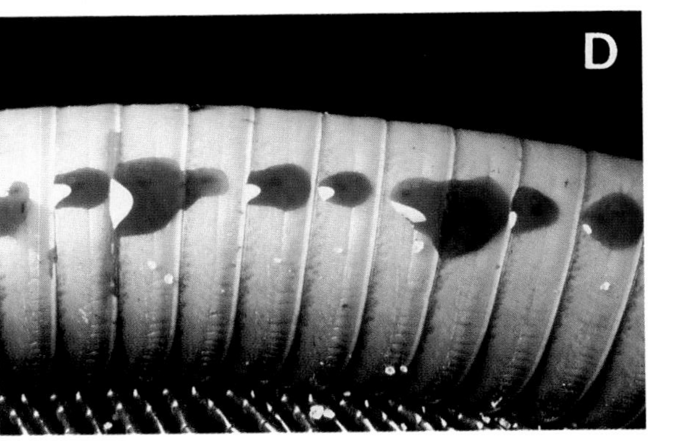

FIGURE 1 Arthropod chemical defenses. (*A*) Blister beetle (*Lytta polita*) reflex-bleeding from the knee joints; the blood contains cantharidin, one of the first insect toxins characterized (4). (*B*) Chrysomelid beetle larva (*Plagiodera versicolora*) emitting droplets of defensive secretion from its segmental glands; the fluid contains methylcyclopentanoid terpenes (chrysomelidial, plagiolactone) (5). (*C*) Whip scorpion (*Mastigoproctus giganteus*) discharging an aimed jet of spray in response to pinching of an appendage with forceps; the spray contains 84% acetic acid (6). (*D*) Millipede (*Narceus gordanus*) discharging its benzoquinone-containing defensive secretion from segmental glands (7). (*Continued overleaf.*)

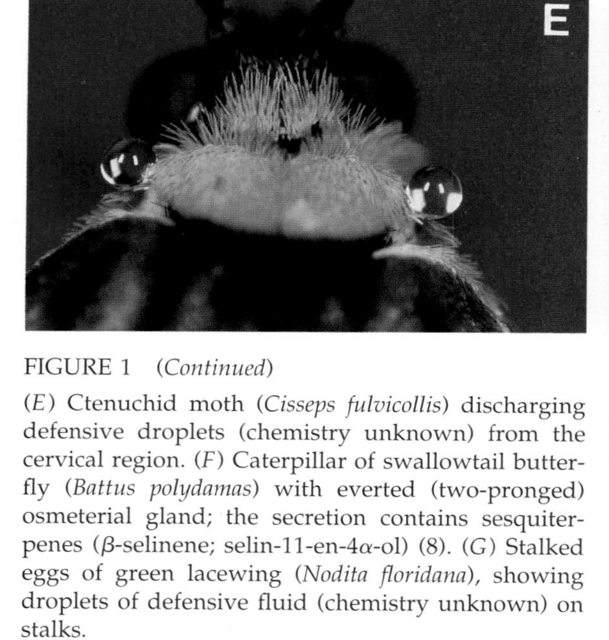

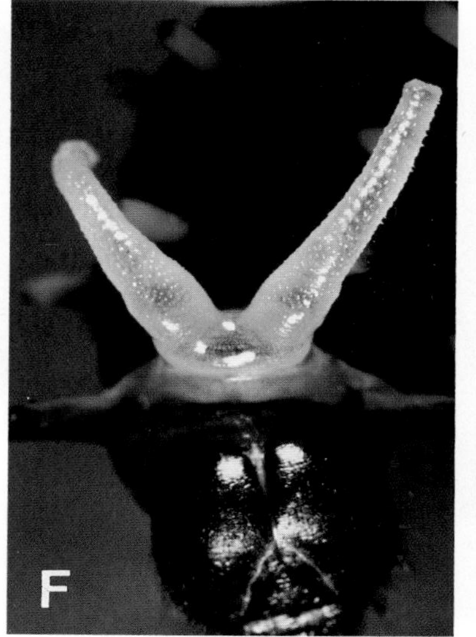

FIGURE 1 (*Continued*)

(*E*) Ctenuchid moth (*Cisseps fulvicollis*) discharging defensive droplets (chemistry unknown) from the cervical region. (*F*) Caterpillar of swallowtail butterfly (*Battus polydamas*) with everted (two-pronged) osmeterial gland; the secretion contains sesquiterpenes (β-selinene; selin-11-en-4α-ol) (8). (*G*) Stalked eggs of green lacewing (*Nodita floridana*), showing droplets of defensive fluid (chemistry unknown) on stalks.

dominant role in the defensive chemistry of dytiscid beetles, and of the large water bug, *Abedus herberti.* This bug discharges a mixture of pregnanes in which desoxycorticosterone is the main component (12). A family of much more highly functionalized steroids, the lucibufagins, serves to render some species of firefly (lampyrid beetles) unpalatable to predatory spiders and birds (13–15). These cardiotonic steroids are closely related to the bufadienolides, whose only known occurrence among animals is in the poison glands of certain toads (16). The discovery

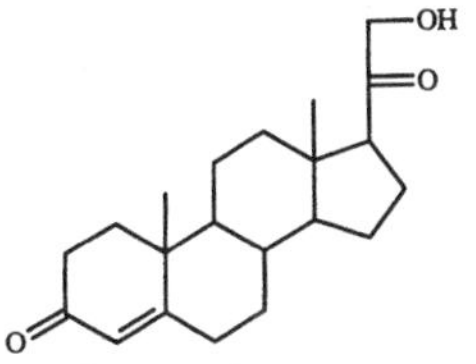

Desoxycorticosterone

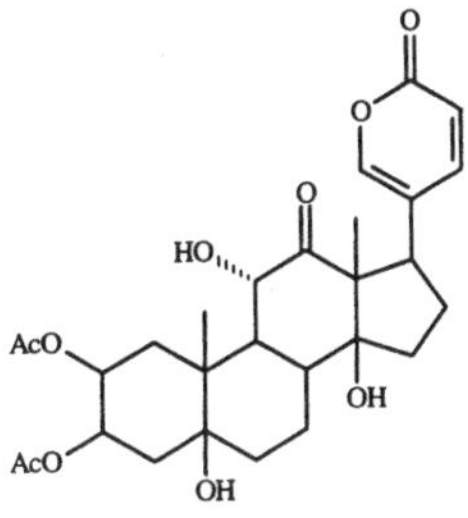

Lucibufagin C

that the lucibufagins also show antiviral properties (17) has prompted us to seek a technique for joining a preformed α-pyrone nucleus to a steroidal framework, since up to now there has been no general, convenient synthetic route to these steroidal pyrones. Our search has recently met with success (18), using the Pd^0-promoted coupling of 5-trimethylstannyl-2-pyrone with an enol triflate (Eq. **1**).

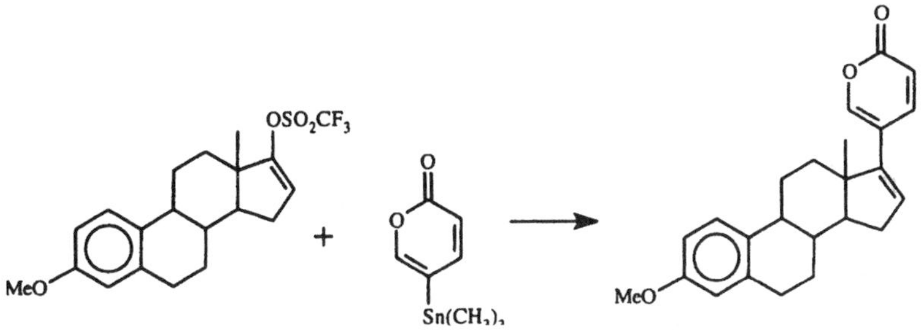

[1]

This direct synthesis of steroidal pyrones should make a variety of structures related to the lucibufagins (as well as to the toad-derived bufadienolides) readily available for biological investigation for the first time. How the insects themselves manage to obtain their defensive pregnanes and steroidal pyrones remains a mystery, since insects are generally considered to lack the enzymatic machinery essential for steroid biosynthesis (19). In fact, we do not yet know whether these insect defensive steroids are produced *de novo* or whether they are derived directly or indirectly from a dietary source; this is a subject which merits further study.

Perhaps the most interesting arthropodan defensive compounds from the point of view of structural diversity are the alkaloids. While alkaloids had long been believed to arise only as a consequence of plant secondary metabolism, it has become apparent over the last few decades that arthropods are both prolific and innovative alkaloid chemists. The millipede *Polyzonium rosalbum*, once thought to secrete camphor (20), in fact gives off a camphoraceous/earthy aroma produced by the spirocyclic isoprenoid imine polyzonimine (21).

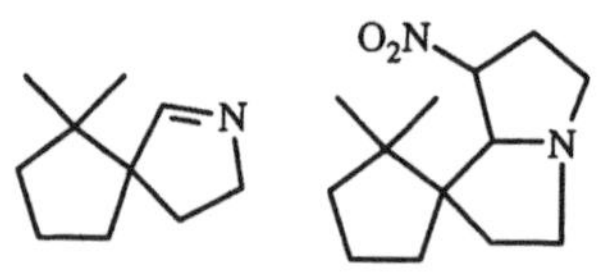

O_2N N N

Polyzonimine   Nitropolyzonamine

The biosynthesis of this imine and its congener, nitropolyzonamine (22), would be another challenging area for future exploration.

We have recently characterized the heptacyclic alkaloid chilocorine from the ladybird (coccinellid) beetle *Chilocorus cacti* (23). In spite of its superficial complexity, this structure is easily dissected into two tricyclic moieties, **A** and **B**, each of which can be regarded as an acetogenin which has been elaborated from a straight chain of 13 carbon atoms stitched together at three points by a trivalent nitrogen atom.

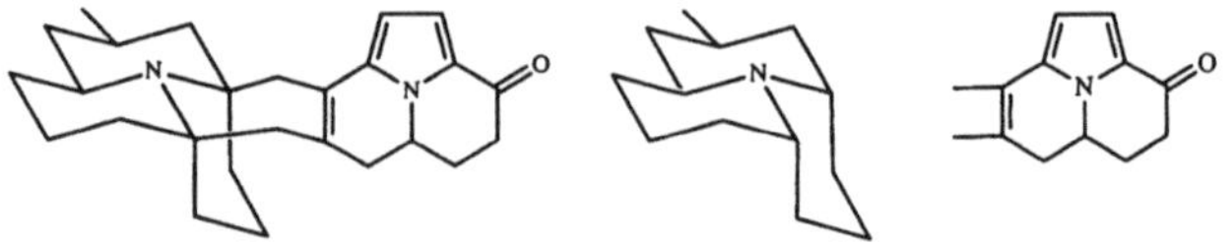

N N O N N O

Chilocorine   **A**   **B**

It is intriguing to note that while the azaphenalene skeleton of part-structure **A** is, in fact, commonly found among coccinellid alkaloids (24), the azaacenaphthylene skeleton of **B** is known, again combined with fragment **A**, only from the recently described hexacyclic alkaloid exochomine (25). Since pyrroles are not basic, we would expect tricyclic compounds resembling **B** to have been missed in a conventional alkaloid isolation scheme, and we anticipate that a targeted search for these novel pyrroles may well turn up additional examples of this otherwise unknown ring system.

Aside from yielding the most complex insect alkaloids so far characterized, coccinellid beetles are sources of a wide array of structural types.

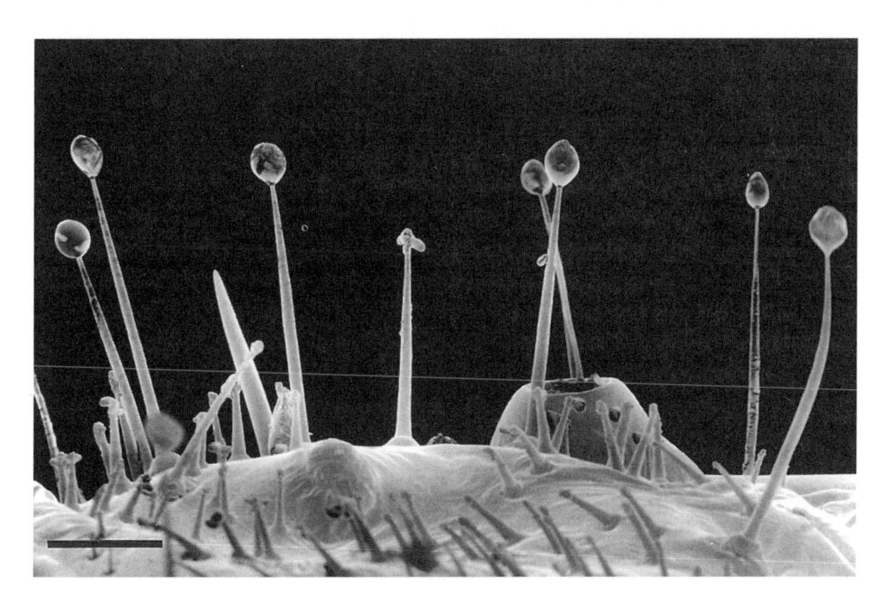

FIGURE 2 Glandular hairs on surface of pupa of the Mexican bean beetle, *Epilachna varivestis.* The secretory droplets contain such azamacrolides as epilachnene (27). (Bar = 0.1 mm.)

Perhaps the star performer in this arena is the Mexican bean beetle, *Epilachna varivestis.* The adult produces a defensive alkaloid cocktail containing more than a dozen pyrrolidines, piperidines, an azabicyclo[3.3.1]nonane, and azaphenalenes (26). Interestingly, the pupa of this beetle, which is densely covered with glandular hairs, secretes an entirely different group of defensive alkaloids, the azamacrolides, which function as highly effective ant repellents (Figure 2). We have described these unique macrocyclic compounds, of which the most important example is epilachnene, in a recent "advertisement" (27). They, too, appear to be constructed from a fatty acid precursor to which a basic nitrogen has been joined. We have carried out a few exploratory biosynthetic experiments with larvae of *E. varivestis* to determine the possible sources of both the 14-carbon straight chain and the ethanolamine moieties of epilachnene (28). Our results are summarized below:

CO_2H O   NH   HO CO_2H NH_2

Oleic acid → Epilachnene ← L-Serine

9,10-Dideuteriooleic acid was shown to be converted to dideuterioepilachnene by the larvae (presumably after two chain-shortening β-oxidations), confirming our expectation that the long carbon chain can be derived from an appropriate fatty acid. Both ^{2}H- and ^{15}N-labeled L-serine are incorporated into the alkaloid as well, accounting for the origin of the ethanolamine unit (28). The unique step in this scheme, as in the most plausible biosynthetic sequences for practically all the coccinellid beetle alkaloids so far characterized, is the joining of a nitrogen substituent to a fatty acid chain. A variety of intriguing mechanistic hypotheses might be put forward to rationalize this process, which does not seem to have any close biochemical precedent. Once more, additional experimental work will be needed to clarify what it is that the beetles actually do, but it seems likely that our understanding of secondary metabolism will make at least a small leap forward as the result of a mechanistic study of this novel carbon–nitrogen bond-forming process.

SPIDER VENOMS

In addition to their widespread exploitation of organic compounds for defensive purposes, arthropod species often use offensive chemical weaponry. Many spiders possess venoms capable of paralyzing their prey, and consequently spider venoms have become a popular hunting ground for neurotoxins of potential neurochemical and neurotherapeutic utility (29). We joined forces with colleagues at Cambridge NeuroScience Research (CNS), Inc., to pursue several problems in this area. In one of these, we studied the venom of *Dolomedes okefinokensis*, a "fishing spider" capable of immobilizing vertebrate prey. The venom of this spider is a typically complex mixture, as revealed by its high-performance liquid chromatogram, which shows the presence of at least several dozen components. Nevertheless, guided by a microscale neurochemical bioassay (30), Kazumi Kobayashi and her CNS colleagues were able to isolate a reversible L- and R-type voltage sensitive calcium channel blocker that appeared to be of interest. We established that this compound, designated CNS 2103, is the 4-hydroxyindole-3-acetic acid amide of a long chain polyamine. The structure, based on ultraviolet absorption and ^{1}H-NMR spectroscopic data, along with conventional and tandem mass spectrometry, was determined to be that given below.

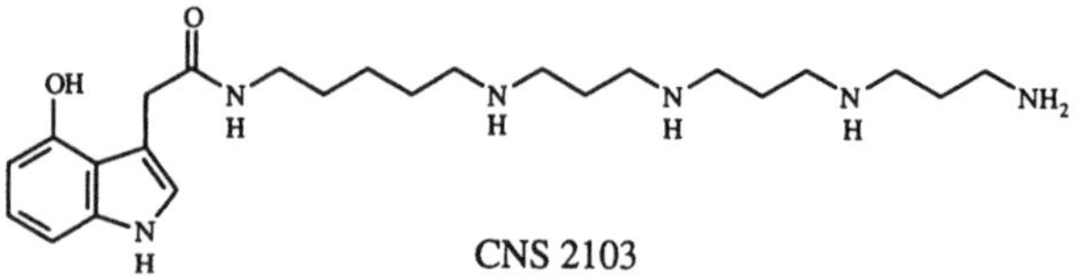

CNS 2103

To confirm this structure, we devised a synthetic route to CNS 2103 which has the virtue of being easily modified to give access to a variety of unnatural analogs of the spider neurotoxin as well (31).

It is of particular interest that polyamines closely related to CNS 2103 have been found not only in other spider species (29) but also in the venom of the solitary digger wasp *Philanthus triangulum* (32). The similarity of these wasp and spider neurotoxins provides a notable example of convergence in the evolution of secondary metabolites aimed at a common target.

Polyamines are readily and reversibly protonated to give water-soluble polycations, and it seems likely that their mode of action is related to this capability (33). In our collaborative work on another venom constituent, HF-7, isolated from the funnel-web spider *Hololena curta*, we encountered an entirely differently constituted blocker of non-*N*-methyl-D-aspartate-glutamate-sensitive calcium channels. This compound turned out to be complementary to the cation-forming polyamines, in the sense that it can exist as a mono- or di-anion. Because of its unusual and entirely unanticipated structure, HF-7 proved more difficult to characterize than CNS 2103. The ultraviolet absorption spectrum of HF-7 pointed to the presence of a guanine chromophore. We had no success in obtaining mass spectroscopic data, however, until we resorted to negative ion fast atom bombardment (FAB) mass spectrometry, which established a molecular weight of 631. Loss of 80 atomic mass units (SO_3) from the m/z 630 parent negative ion gave rise to a base peak corresponding to the molecular formula $C_{18}H_{24}N_5O_{13}S$ (m/z obs. 550.1036; calc. 550.1091), implying the composition $C_{18}H_{25}N_5O_{16}S_2$ for HF-7 itself.

On the basis of a number of two-dimensional NMR experiments, we were finally able to conclude that HF-7 is an acetylated bis(sulfate ester) of a guanosine fucopyranoside, although the exact position of the sulfate groups, the point of attachment of the fucose to the ribose ring, and the absolute configuration of the fucose moiety remained uncertain (34). Because of the very limited supply of this material, and because a family of compounds closely related to the natural product should prove useful in studying structure–activity relationships, we set out to synthesize a set of candidate guanosine fucopyranosides of clearly defined structure and stereochemistry. This synthetic effort has enabled us to characterize HF-7 itself by direct comparison of the natural product with several unambiguously constructed reference compounds (35). As a result of this work, the structure of HF-7 shown below is firmly established.

The occurrence of sulfated nucleoside glycosides in spider venoms, or for that matter in any other natural source, does not seem to have been previously noted. We look forward to the possibility that this novel

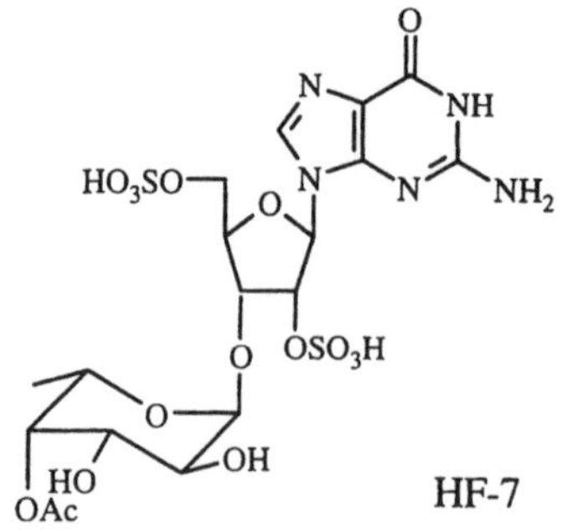

HF-7

arachnid metabolite will prove to be the forerunner of a significant group of anionic neuroactive agents.

CHEMICAL CONTRIBUTIONS TO ARTHROPOD DOMINANCE

In this discussion, we have restricted ourselves to the consideration of only a few examples of arthropod chemistry. From these alone, it is evident that insects synthesize defensive compounds by using all of the major biosynthetic pathways, producing acetogenins, simple aromatics and quinones, isoprenoids, and alkaloids. In addition, some of the millipedes, coccinellid beetles, and spiders we have studied utilize biosynthetic pathways that have yet to be characterized.

While arthropod secretions are often the result of *de novo* syntheses, there are also many instances in which insects pursue an alternative defensive strategy: the sequestration of ready-made defensive compounds from plant or even animal sources. A simple example is provided by larvae of the Australian sawfly *Pseudoperga guerini* (Figure 3), which store the defensive terpenes from ingested eucalyptus, segregating them in a specialized sac, and then regurgitating the mixture in response to attack (36). It is interesting to compare the arthropods' ability to acquire useful natural products with our own species' long history of searching for compounds in nature that can be put to use in a variety of contexts, which has resulted in discoveries of compounds as broadly important as the pyrethroids, quinine, digitoxin, penicillin, the avermectins, and taxol. Although our own searching and screening techniques may well be of unparalleled sophistication, our activities as "chemical prospectors" are certainly not entirely without antecedent.

The preceding falls far short of conveying a true impression of the chemical skills of arthropods. Excluded from our discussion are the diverse signaling agents that mediate such vital insectan functions as food location, mate attraction, social bonding, and alarm communication. Other contributors to this colloquium address some of these topics. While

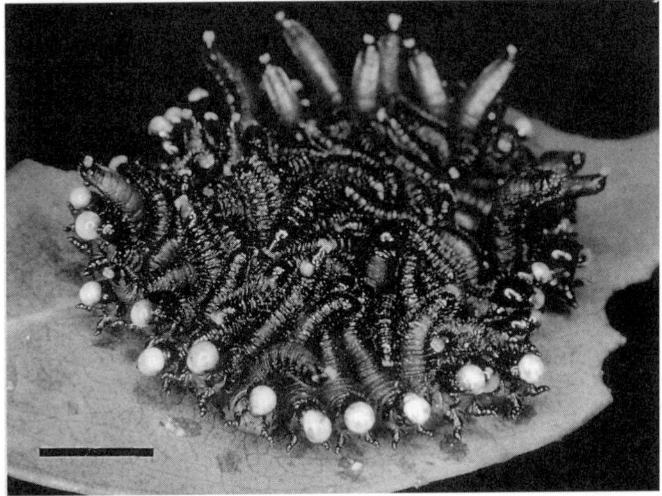

FIGURE 3 Australian sawfly (*Pseudoperga guerini*). (*Top*) Female, guarding clutch of larvae recently emerged from her eggs. At this stage, the larvae are unable to fend for themselves when attacked. (*Bottom*) Older larvae, responding to disturbance by regurgitating droplets of oil from ingested eucalyptus. The droplets are potently deterrent to predators. The newly emerged larvae on left have not as yet accumulated sufficient dietary oil for defense. [Bars = 0.5 cm (*Left*), 1 cm (*Right*).]

insects are by no means unique in having evolved such functions, they may well be the group of animals that took chemical performance in the service of the functions to its most sophisticated expression. The adage "better living through chemistry" may indeed be more applicable to insects than to the industrial giant that coined it (37).

What we know already about insect chemistry is tantalizing, but there can be no question that the best is yet to come. Only a tiny percentage of insects have so far been subject to even the most cursory chemical study. There is no telling what, in the line of molecular novelty and chemical-ecological ingenuity, the remainder might have to offer.

SUMMARY

Studies of arthropod defensive chemistry continue to bring to light novel structures and unanticipated biosynthetic capabilities. Insect alkaloids, such as the heptacyclic acetogenin chilocorine and the azamacrolides, exemplify both of these aspects of arthropod chemistry. Spider venoms are proving to be rich sources of neuroactive components of potential medical interest. The venom of a fishing spider, *Dolomedes okefinokensis*, has yielded a polyamine which reversibly blocks L- and R-type voltage-sensitive calcium channels. Most recently, we have characterized, from the funnel-web spider *Hololena curta*, a sulfated nucleoside glycoside which serves as a reversible blocker of glutamate-sensitive calcium channels. The ability to synthesize or acquire an extremely diverse array of compounds for defense, offense, and communication appears to have contributed significantly to the dominant position that insects and other arthropods have attained.

The support of our research on insect-related chemistry by National Institutes of Health Grants AI12020 and AI2908, National Science Foundation Grant MCB-9221084, and by Hatch Project Grants NY(C)-191424 and NY(C)-191425, as well as by the Schering–Plough Research Institute, the Merck Research Laboratories, and Cambridge NeuroScience, Inc., is gratefully acknowledged.

REFERENCES

1. Erwin, T. L. (1983) in *Tropical Rain Forest: Ecology and Management*, eds. Sutton, S. L., Whitmore, T. C. & Chadwick, A. C. (Blackwell, Edinburgh), pp. 59–75.
2. Wilson, E. O. (1988) in *Biodiversity*, ed. Wilson, E. O. (Natl. Acad. Press, Washington, DC), pp. 3–18.
3. Eisner, T. & Wilson, E. O. (1977) in *The Insects: Readings from Scientific American*, eds. Eisner, T. & Wilson, E. O. (Freeman, San Francisco), pp. 3–15.
4. Carrel, J. E. & Eisner, T. (1974) *Science* **183,** 755–757.
5. Meinwald, J., Jones, T. H., Eisner, T. & Hicks, K. (1977) *Proc. Natl. Acad. Sci. USA* **74,** 2189–2193.

6. Eisner, T., Meinwald, J., Monro, A. & Ghent, R. (1961) *J. Insect Physiol.* **6,** 272–298.
7. Eisner, T., Alsop, D., Hicks, K. & Meinwald, J. (1978) *Handb. Exp. Pharmacol.* **48,** 41–72.
8. Eisner, T., Kluge, A. F., Ikeda, M. I., Meinwald, Y. C. & Meinwald, J. (1971) *J. Insect Physiol.* **17,** 245–250.
9. Attygalle, A. B., Smedley, S. R., Meinwald, J. & Eisner, T. (1993) *J. Chem. Ecol.* **19,** 2089–2104.
10. Eisner, T., Eisner, H. E., Hurst, J. J., Kafatos, F. C. & Meinwald, J. (1963) *Science* **139,** 1218–1220.
11. Jones, T. H., Conner, W. E., Meinwald, J., Eisner, H. E. & Eisner, T. (1976) *J. Chem. Ecol.* **2,** 421–429.
12. Lokensgard, J., Smith, R. L., Eisner, T. & Meinwald, J. (1993) *Experientia* **49,** 175–176.
13. Eisner, T., Wiemer, D. F., Haynes, L. W. & Meinwald, J. (1978) *Proc. Natl. Acad. Sci. USA* **75,** 905–908.
14. Meinwald, J., Wiemer, D. F. & Eisner, T. (1979) *J. Am. Chem. Soc.* **101,** 3055–3060.
15. Goetz, M., Wiemer, D. F., Haynes, L. W., Meinwald, J. & Eisner, T. (1979) *Helv. Chim. Acta* **62,** 1396–1400.
16. Nakanishi, K. (1974) in *Natural Products Chemistry*, eds. Nakanishi, K., Goto, T., Itô, S., Natori, S. & Nozoe, S. (Academic, New York), Vol. 1, p. 469.
17. Wilson, G. R. & Rinehart, K. L. (1989) U.S. Patent 4,847,246.
18. Liu, Z. (1994) Dissertation (Cornell Univ., Ithaca, NY).
19. Nakanishi, K. (1974) in *Natural Products Chemistry*, eds. Nakanishi, K., Goto, T., Itô, S., Natori, S. & Nozoe, S. (Academic, New York), Vol. 1, p. 526.
20. Cook, O. F. (1900) *Science* **12,** 516–521.
21. Smolanoff, J., Kluge, A. F., Meinwald, J., McPhail, A. T., Miller, R. W., Hicks, K. & Eisner, T. (1975) *Science* **188,** 734–736.
22. Meinwald, J., Smolanoff, J., McPhail, A. T., Miller, R. W., Eisner, T. & Hicks, K. (1975) *Tetrahedron Lett.*, 2367–2370.
23. McCormick, K. D., Attygalle, A. B., Xu, S.-C., Svatos, A., Meinwald, J., Houck, M. A., Blankespoor, C. L. & Eisner, T. (1994) *Tetrahedron* **50,** 2365–2372.
24. Ayer, W. A. & Browne, L. M. (1977) *Heterocycles* **7,** 685–707.
25. Timmermans, M., Braekman, J.-C., Daloze, D., Pasteels, J. M., Merlin, J. & Declercq, J.-P. (1992) *Tetrahedron Lett.* **33,** 1281–1284.
26. Attygalle, A. B., Xu, S.-C., McCormick, K. D. & Meinwald, J. (1993) *Tetrahedron* **49,** 9333–9342.
27. Attygalle, A. B., McCormick, K. D., Blankespoor, C. L., Eisner, T. & Meinwald, J. (1993) *Proc. Natl. Acad. Sci. USA* **90,** 5204–5208.
28. Attygalle, A. B., Blankespoor, C. L., Eisner, T. & Meinwald, J. (1994) *Proc. Natl. Acad. Sci. USA* **91,** 12790–12793.
29. McCormick, K. D. & Meinwald, J. (1993) *J. Chem. Ecol.* **19,** 2411–2451.
30. Kobayashi, K., Fischer, J. B., Knapp, A. G., Margolin, L., Daly, D., Reddy, N. L., Roach, B., McCormick, K. D., Meinwald, J. & Goldin, S. M. (1992) *Soc. Neurosci. Abstr.* **18,** 10.
31. McCormick, K. D., Kobayashi, K., Goldin, S. M., Reddy, N. L. & Meinwald, J. (1993) *Tetrahedron* **49,** 11155–11168.
32. Nakanishi, K., Goodnow, R., Konno, K., Niwa, M., Bukownik, R., Kallimopoulos, T. A., Usherwood, P., Eldefrawi, A. T. & Eldefrawi, M. E. (1990) *Pure Appl. Chem.* **62,** 1223–1230.
33. Nakanishi, K., Choi, S.-K., Hwang, D., Lerro, K., Oriando, M., Kalivretenos,

A. G., Eldefrawi, A., Eldefrawi, M. & Usherwood, P. N. R. (1994) *Pure Appl. Chem.* **63,** 671–678.
34. McCormick, K. D. (1993) Dissertation (Cornell Univ., Ithaca, NY).
35. McCormick, J. (1994) Dissertation (Cornell Univ., Ithaca, NY).
36. Morrow, P. A., Bellas, T. E. & Eisner, T. (1976) *Oecologia* **24,** 193–206.
37. Hounshell, D. A. & Smith, J. K., Jr. (1988) *Science and Corporate Strategy: DuPont R&D 1902–1980* (Cambridge Univ. Press, Cambridge, U.K.), p. 221.

The Chemistry of Social Regulation: Multicomponent Signals in Ant Societies

BERT HÖLLDOBLER

The impressive diversity and ecological dominance of ant societies are in large part due to their efficient social organization and the underlying communication system. The functional division into reproductive and sterile castes, the cooperation in rearing the young, gathering food, defending the nest, exploring new foraging grounds, establishing territorial borders, and discriminating and excluding foreigners from the society are regulated by the precise transmission of social signals in time and space.

Probably the best-studied communication behavior in ants is chemical communication, but other sensory modalities, such as mechanical cues, also play an important role in the formation of multicomponent signals in ant communication. Chemical releasers are produced in a variety of exocrine glands, and considerable progress has been made in chemically identifying many of these glandular secretions (for reviews see refs. 1 and 2). In this essay I will not emphasize, however, the natural product chemistry of ant pheromones, but rather concentrate on the proposition that communication in ant societies is often based on multicomponent signals, on nested levels of variation in chemical and other cues, which feature both anonymous and specific characteristics (3).

Bert Hölldobler is professor of zoology and chair of the Department of Behavioral Physiology and Sociobiology at the Theodor-Boveri-Institut für Biowissenschaften der Universität, Würzburg, Germany.

PHEROMONE BLENDS

Single exocrine glands usually produce mixtures of substances. The Dufour's gland secretions of the carpenter ant *Camponotus ligniperda*, for example, include at least 41 compounds (4), and the mandibular glands of the weaver ant *Oecophylla longinoda* contain over 30 compounds, in colony-specific proportions (5, 6). The identification of these components has mostly out-paced an understanding of their function, but in a few cases we begin to realize that the blends are part of the complex behavior-releasing key stimulus. In *Oecophylla*, for example, the mixture of mandibular gland secretions appears to regulate a temporal sequence of orientation and aggressive reactions, as different components diffuse outward from the point of origin (7). Vander Meer and his collaborators investigated the recruitment and trail following behavior of the fire ant (*Solenopsis* spp.), and they obtained experimental results which suggest that these behaviors are released by a complex blend of compounds derived from the Dufour's gland (8, 9). The principal components for trail orientation are two α-farnesenes, two homofarnesenes, and a still unidentified component which releases attraction behavior. Oddly, these substances remain inactive unless the ants have been induced by yet another, still-unidentified, component in the glandular secretions. The two substances responsible for attraction and inducing require about 250 times the relative concentration of the orientation pheromone. These recent findings can be conclusively related to previous observations that the more desirable the food find, the more intense is the trail laid by the recruiter ants (2). High quantities of discharged trail pheromone provide a sufficient amount of initial attraction and inducer pheromones to get the recruitment process started. Once "turned on" by these signals, the ants also follow trails consisting of relatively small amounts of orientation pheromones. A similar, although less complicated, effect of pheromone blends releasing a behavioral response was delineated by Attygalle and Morgan (10) in *Tetramorium caespitum*. This myrmicine lays trails composed of two pyrazines; workers respond maximally to a blend with a weight ratio of 3:7 of the two substances. However, such pheromone blending has not always arisen in evolution: in eight species of *Myrmica* trail following is released by the same single compound, 3-ethyl-2,5-dimethylpyrazine (11). The absence of species specificity in chemical recruitment trails has been reported in a number of ant species (2). However, cross-species trail-following does not necessarily mean identical trail pheromones. For example, in two closely related species of *Aphaenogaster*, *A. cockerelli* follows only its own trail, whereas *A. albisetosus* responds to the trails drawn with poison gland secretions of both species (12). A recent chemical analysis of poison gland contents of

Aphaenogaster conducted by D. Morgan and his collaborators and subsequent behavioral tests of the identified compounds revealed that the main recruitment pheromone of *A. cockerelli* is (*R*)-(+)-1-phenylethanol, and that of *A. albisetosus* is 4-methyl-3-heptanone (41). But *A. cockerelli* poison gland secretions also contain 4-methyl-3-heptanone and 4-methyl-3-heptanol; *A. cockerelli* workers do not, however, respond to these latter secretions with trail-following behavior. On the other hand, *A. albisetosus* respond to the 4-methyl-3-heptanone in the *A. cockerelli* secretions but not to the phenylethanol. Incidentally, the latter substance was not previously known to be an ant pheromone and is very unusual for poison gland contents. Thus, although we find cross-specific responses in *A. albisetosus*, this species reacts to a different component in the *A. cockerelli* trail than do *A. cockerelli* workers themselves.

In addition to pheromone blends in a single exocrine gland, multicomponent signals can derive from multisource systems. In such systems various compounds are released from multiple glandular sources. The substances may serve the same essential functions, but often the roles are different. In the harvester ant *Pogonomyrmex badius*, for example, the recruitment pheromone is voided from the poison gland, whereas the long-lasting homing pheromones originate at least in part in the Dufour's gland (13). Further investigations of this multisource system in the genus *Pogonomyrmex* have revealed that the recruitment signal is, as far as we know, invariant among several sympatric *Pogonomyrmex* species, whereas the Dufour's gland secretions contain species-specific mixtures of hydrocarbons (14, 15). Field and laboratory investigations suggest that in the partitioning of foraging areas among sympatric species of *Pogonomyrmex* both the short-lived anonymous recruitment signals and the more persistent species-specific Dufour's gland secretions are involved. The latter appear to mark the trunk routes, which also bear colony-specific markers, the origins of which are not yet known. A similar situation has been observed in ants of the genus *Myrmica*, which produce relatively anonymous recruitment signals originating in the poison gland and species-specific mixtures of hydrocarbons in the Dufour's glands that are used as home range markers (10, 11).

Many ponerine ant species conduct predatory raids on termites and other arthropods, and generally these are organized by powerful trail pheromones which are often composed of secretions from several glands. In group-raiding *Leptogenys* species one component originates from the poison gland, but a second orientation pheromone, (3*R*,4*S*)-4-methyl-3-heptanol, derives from the pygidial gland (16, 17). We recently discovered an identical situation in *Megaponera foetens* (18). In both cases the poison gland secretions have a stronger orientation effect, while the pygidial gland secretions serve as the major recruitment signal. In several species of the legionary genus *Onychomyrmex*, both group raiding and colony emigrations are organized by trails laid with a sternal gland. However, chemical orientation appears to be supplemented by homing signals deposited from a basitarsal gland in the hindlegs (19).

Homing signals are often colony specific. A still finer level of specificity has recently been demonstrated among individual colony members, a surprising finding, given the prevailing view that individual differentiation among social insect workers is weak. Individual-specific orientation trails have been discovered in the ants *Pachycondyla tesserinoda* and *Leptothorax affinis* (20, 21), among others. The source of these highly specific markers and how they are chemically composed are not yet known.

In general, specificity in a multicomponent signal seems to be a form of modulation. Assuming that modulatory functions presuppose the existence of the behavior being modulated, a possible evolutionary route to signal specificity can be proposed (3).

The production of simple semiochemicals, releasing simple, anonymous reactions, is subject to the inevitable imprecision of all biosynthetic processes. The resulting degree of variation may well be perceptible to the receiver's sensory system, but it will ordinarily have no effect on the response to the signal. However, should an adaptive advantage happen to correlate with any of the available variants, selection will favor individuals which respond differently on the basis of these specific characteristics—i.e., modulation of the original response. Take as an example undecane, the anonymous alarm signal of many species of the subfamily Formicinae. It is usually the most abundant product in the formicine Dufour's glands. However, other hydrocarbons are also present and the total mixture is often species specific (11). Thus, during alarm behavior undecane will be discharged together with a blend of other hydrocarbons. If, say, genetically similar colony members tend to produce similar hydrocarbon patterns, the signal may come to be modulated by this added specificity, informing workers whether nestmates or aliens are releasing the alarm. Once the presence and/or proportions of additional components significantly affect the response to the basic releaser in an adaptive manner, selection is expected to improve their distinctiveness and stereotyping.

MODULATION AND RITUALIZATION OF MULTICOMPONENT SIGNALS

This evolutionary process by which a phenotypic trait is altered to serve more efficiently as a signal is called *ritualization*. Commonly, the process begins when some movement, anatomical feature, or physiological or biochemical trait that is functional in another context acquires a secondary value as a signal.

The evolution of specificity in a multicomponent signal described above can be interpreted as chemical ritualization, whereby increasingly functional specificity could derive from the biochemical "noise" in an ancestral anonymous signal and ritualization of specific variation is likewise possible with chemicals that initially were uninvolved in communication. For example, species-specific trail pheromones from the poison glands of myrmicine ants are generally the metabolic by-products of venom synthesis (11), while the Dufour's gland hydrocarbons of formicines sprayed together with formic acid may enhance its spread and penetration. In fact, the evolutionary process of ritualization appears to have played an important role in the evolution of diverse modes of communication behavior in ant societies and is closely connected with the evolution of modulatory communication. Communication in complex social systems is not always characterized by a deterministic releasing process but sometimes plays a more subtle role. For example, in a group of ant workers certain communication signals suffice to adjust the behavior of group mates towards one another. These signals have the effect of shifting the probability for the performance of other behavioral acts, but they do not elicit particular behavioral responses. We have called this kind of communication system "modulatory communication" (22, 23). Modulatory signals are devices for shifting the threshold for the releasing effectiveness of other stimuli, thus enhancing the behavioral response to them. In this sense, the orientation-inducer pheromone in *Solenopsis* may also be called a modulatory signal (9).

In only a few cases has a statistical information analysis of modulatory communication been carried out; circumstantial evidence suggests, however, that it is widespread in insect societies. These more rigorously analyzed cases of modulatory communication concerned situations in which one signal modulates another of a different modality (22, 23). For example, in *A. cockerelli* or *A. albisetosus* a forager, after discovering a prey object too large to be carried or dragged by a single ant, releases poison gland secretion into the air. Nestmates as far away as 2 m are attracted and move toward the source. When a sufficient number of foragers have assembled around the prey, they gang-carry it swiftly to the nest. Time is of the essence, because *Aphaenogaster* must remove food from the scene before formidable mass-recruiting competitors, including fire ants and *Forelius pruinosus*, arrive in large numbers. *Aphaenogaster* workers, in addition to releasing the poison gland pheromone, also regularly stridulate at the prey object. Ants perceiving the substrate-borne signals start to encircle the prey sooner, and they are likely to release the attractive poison gland pheromone earlier. Overall, both the recruitment of workers and the retrieval of the food object are advanced by 1–2 min as a consequence of stridulation (22).

It is conceivable that such rather unspecific modulatory signals obtain more specific significance in the communication process. A striking example is that of the leaf-cutter ant *Atta cephalotes* (24).

Atta workers stridulate when cutting an attractive leaf. Stridulatory vibrations migrate along the body of the leaf-cutting ant and are led into the substrate through the ant's head. We have evidence that the vibrations caused by the stridulation mechanically facilitate the cutting process. We could, in addition, demonstrate that the substrate-borne vibrations not only enhance the chemical recruitment signal, laid with poison gland secretions, but also suffice to attract nestmates to the cutting site. We therefore hypothesize that a motor pattern whose original function might have been to support the cutting process, secondarily became a modulatory signal, and subsequently has further evolved to function as an independent recruitment signal. In fact, Markl (25) has demonstrated that in another behavioral context the stridulatory substrate vibrations in *Atta* serve as stress and rescue signals.

Another striking example of the evolution of multicomponent signals in ant communication is found in the multiple recruitment system of the weaver ants (*Oecophylla*) (26). Workers of this genus utilize no fewer than five recruitment systems: for summoning nestmates to new food sources, to new terrain, for emigration, to territorial defense, and (short range) to territorial intruders. Although the messages differ from one another strongly, they are built out of pheromones from two or three organs—the rectal, sternal, and possibly also the mandibular gland—together with a modest array of stereotyped movements and tactile stimuli. The specificity of each of the recruitment systems comes principally from the combinations of chemical and tactile elements. For example, both recruitment to food and recruitment to territorial defense are guided by pheromones from the rectal gland. Territorial defense is further specified by forward jerking movements which closely resemble maneuvering during actual attack behavior. We have therefore interpreted the signals to be a ritualized version, "liberated" during evolution to serve as a signal when a nestmate is encountered rather than an enemy. When workers recruit nestmates to food, they use a wholly different set of movements. They wave their heads laterally while opening their mandibles. The movement resembles that of food offering and may have derived from that through a ritualization process. Other communicative motor patterns in ants—such as short runs or jerking or wagging motions employed during recruitment communication to summon nestmates to food sources, to nest sites, or to the defense of territories (26)—may in part have evolved from motor displays that originally served as general modulators. They have since been ritualized into specialized signals employed in specific contexts, usually in combination with other signals such as trail or alarm pheromones (27).

NESTMATE RECOGNITION AND EXCLUSION OF FOREIGNERS

In most social insects, interactions between conspecific adults from different colonies are quite aggressive. Such behavior is considered to be adaptive, as workers obtain inclusive fitness benefits from aiding kin and discriminating against non-kin, and nestmates are usually more closely related to one another than to members of neighboring colonies. The semiochemicals involved in recognition at the colony level are simultaneously specific and anonymous. That is, workers are able to discriminate between nestmates and intruders, but they also tend to treat all nestmates as fellow colony members, irrespective of their true relatedness. This anonymity among genetically varied nestmates does not preclude specificity at the within-colony level. Generally, though, it appears that workers encountering one another in the context of territorial defense or nest guarding respond to chemical labels that indicate colony membership, rather than directly indicating kin. Especially species with larger colonies are characterized by a more or less homogeneous recognition signal or "colony odor," specific between colonies but anonymous throughout each colony. The sources of nestmate recognition signals in social insects have recently received a great deal of attention (for partial reviews see refs. 2, 3, 28, and 29), and several investigators paid special attention to possible colony-specific profiles of cuticular hydrocarbons (i.e., refs. 30–33). Obviously, to achieve such a high degree of specificity, such recognition labels must be rather complex, multicomponent signals. However, there exists no conclusive proof yet that the implied colony-specific hydrocarbon patterns serve as nestmate recognition labels, nor is it possible yet to develop a unified concept of the sociobiological foundations of nestmate recognition. The current results indicate that the behavioral mechanisms underlying nestmate recognition vary with the specific social organizations of the societies.

In monogynous (only one queen is present) carpenter ants (*Camponotus* spp.), for example, nestmates are distinguished by chemical labels acquired from a variety of sources, functioning in a hierarchical order of significance (2, 3, 34, 35). Workers removed as pupae from a colony and reared separately, in the absence of queens, are relatively tolerant of one another, but exhibit stronger aggressive behavior toward nonrelatives. Diet differences slightly enhance aggression among separately reared kin. If a queen is present, however, workers attack both unfamiliar kin and non-kin with equal violence, a response which is unaffected by food odors. Cues derived from healthy queens with active ovaries are sufficient to label all workers in experimental colonies, while the workers' own discriminators become more important when their queen is infertile. As mentioned above, recent work suggests that the acquired recognition cues of *Camponotus* spp. workers may consist at least in part of colony-specific relative proportions of cuticular hydrocarbons (30, 31, 33). The queen is by no means the only source of shared extrinsic recognition cues. The discriminators produced by each worker may be transferred among them all, resulting in a "gestalt" or mixed label, as originally proposed by Crozier and Dix (36) and demonstrated, among others, in leptothoracine ants (37). In addition to heritable cues from other workers and/or queens, variation originating in the diet or other environmental differences external to the colony also contributes to nestmate recognition in several ant genera (for review see refs. 2, 3, and 38).

CONCLUSION

The early discovery of such extremely fine-tuned chemical sexual communication as that of the silkmoth *Bombyx mori* (39) encouraged the belief that, among insects, each behavioral response is released by a single chemical substance. By contrast, much greater population and individual variability was attributed to the chemical communication signals produced by vertebrates, particularly mammals, in which pheromones often mediate more interindividual interactions such as individual recognition, dominance ranking, and territorial marking. While the complex chemical composition of mammalian pheromones was examined for functional significance, the same degree of variation observed in an insect pheromone would be ascribed to contamination or biosynthetic "noise." It is now clear that such a double standard was, at best, an oversimplification. Most insect semiochemicals have proven to be complex mixtures, and single-compound pheromones are actually rare (40). In this respect at least, insects and vertebrates do not differ greatly in the sophistication of their chemical communication systems.

SUMMARY

Chemical signals mediating communication in ant societies are usually complex mixtures of substances with considerable variation in molecular composition and in relative proportions of components. Such multicomponent signals can be produced in single exocrine glands, but they can also be composed with secretions from several glands. This variation is often functional, identifying groups or specific actions on a variety of organizational levels. Chemical signals can be further combined with cues from other sensory modalities, such as vibrational or tactile stimuli. These kinds of accessory signals usually serve in modulatory communication, lowering the response threshold in the recipient for the actual releasing stimulus. Comparative studies suggest that modulatory signals evolved through ritualization from actions originally not related to the same behavioral context, and modulatory signals may further evolve to become independent releasing signals.

I thank J. Heinze, M. Lindauer, C. Peeters, F. Roces, and J. Tautz for reading the manuscript. Special thanks are due to my former collaborator N. F. Carlin, with whom several of the concepts concerning anonymity and specificity of chemical signals have been developed in a joint paper (3). Some of the author's work presented in this essay has been made possible by grants from the National Science Foundation and the Leibniz-Prize from the Deutsche Forschungsgemeinschaft.

REFERENCES

1. Wheeler, J. W. & Duffield, R. M. (1988) in *Handbook of Natural Pesticides*, eds. Morgan, E. D. & Mandave, N. B. (CRC, Boca Raton, FL), Vol. 4, pp. 59–206.
2. Hölldobler, B. & Wilson, E. O. (1990) *The Ants* (Belknap Press of Harvard Univ. Press, Cambridge, MA).
3. Hölldobler, B. & Carlin, N. F. (1987) *J. Comp. Physiol. A* **161,** 567–581.
4. Bergström, G. & Löfquist, J. (1972) *Entomol. Scand.* **3,** 225–238.
5. Bradshaw, J. S., Baker, R. & Howse, P. E. (1975) *Nature (London)* **258,** 230–231.
6. Bradshaw, J. S., Baker, R., Howse, P. E. & Higgs, M. D. (1979) *Physiol. Entomol.* **4,** 27–38.
7. Bradshaw, J. S., Baker, R. & Howse, P. E. (1979) *Physiol. Entomol.* **4,** 15–25.
8. Vander Meer, R. K., Alvarez, F. & Lofgren, C. S. (1988) *J. Chem. Ecol.* **14,** 825–838.
9. Vander Meer, R. K., Lofgren, S. & Alvarez, F. M. (1990) *Physiol. Entomol.* **15,** 483–488.
10. Attygalle, A. & Morgan, E. D. (1985) *Adv. Insect Physiol.* **18,** 1–30.
11. Morgan, E. D. (1984) in *Insect Communication*, ed. Lewis, T. (Academic, New York), pp. 169–194.
12. Hölldobler, B., Stanton, R. C. & Markl, H. (1978) *Behav. Ecol. Sociobiol.* **4,** 163–181.
13. Hölldobler, B. & Wilson, E. O. (1970) *Psyche* **77,** 385–399.
14. Regnier, F., Nieh, M. & Hölldobler, B. (1973) *J. Insect Physiol.* **19,** 981–992.
15. Hölldobler, B. (1986) in *Information Processing in Animals*, ed. Lindauer, M. (Fischer, Stuttgart), Vol. 3, pp. 25–70.
16. Maschwitz, U. & Schönegge, P. (1977) *Naturwissenschaften* **64,** 589–590.
17. Attygalle, A. B., Vostrowsky, O., Bestmann, H. J., Steghaus-Kovac, S. & Maschwitz, U. (1988) *Naturwissenschaften* **75,** 315–317.
18. Hölldobler, B., Braun, U., Gronenberg, W., Kirchner, W. & Peeters, C.J. (1994) *Insect Physiol.* **40,** 585–593.
19. Hölldobler, B. & Palmer, J. M. (1989) *Naturwissenschaften* **76,** 385–386.
20. Jessen, K. & Maschwitz, U. (1985) *Naturwissenschaften* **73,** 549–550.

21. Maschwitz, U., Lenz, A. & Buschinger, A. (1986) *Experientia* **42,** 1173–1174.
22. Markl, H. & Hölldobler, B. (1978) *Behav. Ecol. Sociobiol.* **4,** 183–216.
23. Markl, H. (1985) in *Experimental Behavioral Ecology and Sociobiology,* eds. Hölldobler, B. & Lindauer, M. (Fischer, Stuttgart), pp. 163–194.
24. Roces, F., Tautz, J. & Hölldobler, B. (1993) *Naturwissenschaften* **80,** 521–524.
25. Markl, H. (1965) *Science* **149,** 1392–1393.
26. Hölldobler, B. & Wilson, E. O. (1978) *Behav. Ecol. Sociobiol.* **3,** 19–60.
27. Hölldobler, B. (1978) *Adv. Study Behav.* **8,** 75–115.
28. Fletcher, D. J. C. & Michener, C. D., eds. (1987) *Kin Recognition in Animals* (Wiley, New York).
29. Heppner, P. G., ed. (1991) *Kin Recognition* (Cambridge Univ. Press, New York).
30. Bonavita-Cougourdan, A., Clément, J. L. & Lange, C. (1987) *J. Entomol. Sci.* **22,** 1–10.
31. Morel, L., Vander Meer, R. K. & Lavine, B. K. (1988) *Behav. Ecol. Sociobiol.* **22,** 175–183.
32. Vander Meer, R. K., Saliwanckik, D. & Lavine, B. (1989) *J. Chem. Ecol.* **15,** 2115–2125.
33. Lavine, B. K., Morel, L., Vander Meer, R. K., Guderson, R. W., Han, J. H., Bonanno, A. & Stine, A. (1990) *Chemometrics Intell. Lab. Syst.* **9,** 107–114.
34. Carlin, N. F. & Hölldobler, B. (1986) *Behav. Ecol. Sociobiol.* **19,** 123–134.
35. Carlin, N. F. & Hölldobler, B. (1987) *Behav. Ecol. Sociobiol.* **20,** 209–218.
36. Crozier, R. H. & Dix, M. W. (1979) *Behav. Ecol. Sociobiol.* **4,** 217–224.
37. Stuart, R. J. (1988) *Proc. Natl. Acad. Sci. USA* **85,** 4572–4575.
38. Carlin, N. F. (1989) *Neth. J. Zool.* **39,** 86–100.
39. Schneider, D. (1957) *Z. Vergl. Physiol.* **40,** 8–41.
40. Silverstein, R. M. & Young, J. C. (1976) in *Pest Management with Insect Sex Attractants,* ed. Gould, R. F. (Am. Chem. Soc., Washington, DC), pp. 1–29.
41. Oldham, N.J., Morgan, E.D. & Hölldobler, B. (1994) in *Les Insectes Sociaux,* eds. Lenoir, A., Arnold, G. & Lepage, M. (Université Paris Nord, Paris), p. 486.

The Chemistry of Eavesdropping, Alarm, and Deceit

MARK K. STOWE, TED C. J. TURLINGS, JOHN H. LOUGHRIN, W. JOE LEWIS, AND JAMES H. TUMLINSON

A diverse multitude of arthropods hunt other arthropods as food for themselves or for their progeny. It is becoming increasingly obvious that in many of these systems, chemical signals, or "semiochemicals" (1), are crucial to the hunters' success. These semiochemicals can function in a variety of ways to bring hunters and quarry together. In the simplest systems, arthropod predators or parasitoids are attracted to their prey or hosts by semiochemicals called "kairomones" (2) emitted by the victims. Often the semiochemicals used by parasitoids or predators to locate their hosts or prey function as different types of signals in other communication systems. Thus, there is an overlapping of chemical signals with the hunters intercepting messages "intended for" a different receiver. For example, many parasitoids exploit the pheromonal signals of their hosts during foraging. This strategy, which Vinson (3) has termed "chemical espionage" (see also ref. 4), is effective only when pheromones indicate the location of the life stage of the host that is the

Mark Stowe is a research associate in the Department of Zoology at the University of Florida, Gainesville. Ted Turlings is leader of the Chemical and Behavioral Ecology Research Team in the Institute of Plant Sciences/Applied Entomology at the Swiss Federal Institute of Technology, Zurich. John Loughrin is a post-doctoral scholar in the Department of Entomology at the University of Kentucky, Lexington. W. Joe Lewis is a research entomologist in the Insect Biology and Population Management Research Laboratory of the U.S. Department of Agriculture, Tifton, Georgia. James Tumlinson is research leader in the Insect Attractants, Behavior and Basic Biology Research Laboratory of the U.S. Department of Agriculture, Gainesville, Florida.

target of the parasitoid. When the target life stage does not reveal itself by long-distance pheromonal signals, predators and parasitoids have been forced to adopt other strategies. In some systems parasitoids or predators locate herbivorous prey by exploiting plant signals induced by the herbivores (5–8). Thus, both the plants and the predators or parasitoids benefit from this interaction. In contrast to the foraging predators, some arthropods remain stationary and emit mimetic pheromone signals to attract and capture their prey (9).

During the last decade we have been investigating the chemically mediated foraging behavior of beneficial entomophagous arthropods in an effort to elucidate the factors that guide them to their hosts or prey. Our ultimate goal is to be able to manipulate and control these organisms to increase their effectiveness as biological control agents and thus reduce our dependence on pesticides for control of insect pests in agriculture. As we and our colleagues have learned more about these systems, we have found them to be quite complex in many instances. We have also found a surprising diversity of mechanisms by which these systems operate. Here we briefly survey three categories of chemically mediated predator–prey relationships which we have arbitrarily termed "eavesdropping, alarm, and deceit." Recent reviews (8–13) describe many of these systems in more detail.

EAVESDROPPING

Pheromones, by definition, are chemical signals between two members of the same species. In most instances sex pheromones are highly specific, attracting members of the same species only and not those of closely related species. However, it appears that many predatory and parasitic arthropods are able to intercept the sex pheromone signals of their prey or hosts. Bedard in 1965 (as cited in ref. 14) first reported the attraction of a parasitic wasp, the pteromalid *Tomicobia tibialis* Ashmead, to volatiles produced by males of the bark beetle *Ips paraconfusus* (Le Cônte) boring in ponderosa pine. While the identity of the volatile kairomone in this case has not been determined, it is very likely the sex or aggregating pheromone produced by the beetles. In analogous studies several bark beetle predators have been captured in traps baited with synthetic components of the pheromones of *Ips* and *Dendroctonus* species (14). Also, several hymenopterous parasites of the elm bark beetle, *Scolytus multistriatus* (Marsham), are attracted to components of its pheromone, multistriatin, 4-methyl-3-heptanol, and cubebene and combinations thereof (15).

Corn earworm moth, *Helicoverpa zea*, females emit a blend of hexadecanal and (Z)-7-, (Z)-9-, and (Z)-11-hexadecenal (16) that is a highly specific attractant for *H. zea* males. However, we discovered that field application of synthetic *H. zea* sex pheromone significantly increased rates of parasitization of *H. zea* eggs by naturally occurring female wasps belonging to the chalcid genus *Trichogramma* (17). Similarly, Noldus and van Lenteren (18) found that *Trichogramma evanescens* Westwood females respond to sex pheromones emitted by females of *Pieris brassicae* L. and *Mamestra brassicae* L. in a laboratory olfactometer. These findings came as a surprise because female moths release sex pheromones at night, while the wasps are only active during the day. Subsequent studies support the hypothesis that wasps are able to detect sex pheromone scent which has adsorbed onto the leaf surfaces near calling female moths at night and which is still being released by the leaves the following day. Noldus *et al.* (19) exposed leaves to calling female moths and then tested wasps for their response to exposed and unexposed leaves. Wasps showed a significantly greater response to exposed leaves for as long as 24 hr after exposure.

Lingering sex pheromone scents are useful clues to the general location of moth eggs. Female noctuid moths call in the presence of host plants and usually lay eggs nearby. Because the pheromone scent emanates from a source (leaves) that is only an approximate indication of the location of the target (moth eggs), the wasps (in contrast to male moths) do not fly upwind in pheromone plumes. Instead they respond to pheromone odor by landing on nearby surfaces and then rely on visual and short-range chemical cues to find the eggs themselves.

At present only *Trichogramma* species and the scelionid egg parasitoid *Telenomus remus* (20) are known to respond to moth sex pheromones. However, this phenomenon could be much more widespread and overlooked because these wasps apparently do not fly to point sources and are not caught in pheromone traps.

A different eavesdropping system can be found in the relationship between true bugs (Heteroptera) and some of their parasites and parasitoids (reviewed in ref. 21). The male bugs emit sex pheromones to attract females. In the process they also attract diverse parasitic flies and wasps. Unlike the *Trichogramma* wasps that attack moth eggs, these flies and wasps are attracted to point sources of synthetic pheromone compounds. Sticky traps baited with the synthetic sex pheromone [(*E*)-2-hexenal, benzyl alcohol, linalool, terpenen-4-ol, α-terpineol, and piperitol] (21) of the spined soldier bug *Podisus maculiventris* caught more than 17,800 parasitoids in three seasons—more than 5 times the number of female bugs captured. Two species of tachinid flies, *Euclytia flava* and *Hemyda aurata*, are highly attracted to the soldier bug pheromones and lay eggs, primarily on adult males. Female soldier bugs escape parasitization, except during mating, and consequently have about 25% as many tachinid eggs on their bodies as do males. Interestingly, a "cheater" male strategy has apparently evolved. Soldier bug males are attracted by pheromone produced by other males and wait in close proximity to the pheromone-emitting males while not emitting pheromone themselves. Thus, these "cheater" males attempt to intercept attracted females without being parasitized.

Tiny ceratopogonid flies are also attracted to the pheromone released by calling male bugs. They puncture the bugs' pronotum and engorge on blood, apparently without disturbing the bugs. Some of the ceratopogonid flies are known to be extreme generalists and may be capable of eavesdropping on a wide range of pheromones (21).

Females of the wasp *Telonomus calvus* parasitize the eggs of *P. maculiventris* and *Podisus fretus*. They wait in the vicinity of male bugs releasing pheromone and become phoretic on female bugs that mate with the males (22, 23). *T. calvus* requires eggs less than 12 hr old to successfully develop. Thus, by riding on the female until she oviposits, the female wasps are assured of finding fresh eggs before they are discovered by competing parasitoids or predators (21).

All of the long-range kairomones attractive to parasitoids that have been identified thus far are sex pheromones of the hosts. However, we are probably aware of only a small fraction of the predators and parasites that are eavesdropping on the pheromonal communications of their prey or hosts. While the evolution of individuals that are as inconspicuous as possible to their enemies is favored, it is impossible for a species to completely avoid emitting chemical signals. Thus, pheromones that are important to reproduction or other vital functions, and are good indicators of the presence of a species, are available for predators or parasitoids to exploit.

ALARM

One of the most interesting systems that we have studied involves interactions of herbivorous larvae, the plants on which they feed, and parasitic wasps that attack the larvae. On the one hand, the larvae have evolved to become as inconspicuous as possible to avoid parasitization and predation. However, they must feed to survive and in feeding they damage the plant, and in doing so induce a reaction from the plant. What is most surprising is the plant's reaction. In addition to passively releasing volatile chemicals from their damaged tissues, plants under attack actively produce and release volatile compounds from undamaged as well as damaged tissues. This suggests a plant defensive mechanism to repel invaders since many of the volatiles released by damaged plants have been shown to be insect repellents. However, the wasps are clearly not repelled by these volatile odors released by damaged plants, but, on the contrary, exploit them to find their hosts. Thus, both the plants and the wasps may benefit.

The first indication of the active role of plants in producing volatile chemicals to attract the natural enemies of their herbivorous attackers was found by Dicke, Sabelis, and coworkers (7, 24) in their studies of predatory mites that prey on plant-feeding mites. They found that when herbivorous spider mites feed on lima bean leaves, the plant releases a blend of volatiles that attracts predatory mites. The blend produced differs between plant species and varies depending on the species of spider mite that is attacking the plant. The blends even differ between plant cultivars infested with the same spider mite species, and the predatory mites can detect these differences (25, 26). Artificially damaged leaves are not attractive to the predatory mites.

Behavioral studies in our laboratories on host foraging of *Cotesia marginiventris*, a parasitoid of larvae of several species of noctuid moths, indicated that plants damaged by hosts were the most important source of volatile attractants for the female wasps (6, 27). Removal of hosts and host products, including feces, slightly diminished the attractiveness of the plants, but neither hosts nor feces were as attractive as the damaged plants alone. Collection and analysis of volatiles produced by corn seedlings fed on by beet armyworm, *Spodoptera exigua*, larvae overnight showed that the corn was producing both short chain compounds, like the 6-carbon aldehydes, alcohols, and esters normally associated with green leafy odors, as well as indole and several terpenes and sesquiterpenes (28) (Figure 1). However, when larvae were allowed to feed on fresh seedlings for only 2 hr, during which time volatiles were collected and analyzed, only the green leafy compounds were found. Subsequent tests showed that the plants only begin producing the terpenes several hours after damage. Furthermore, the green leafy volatiles were only produced during active caterpillar feeding. When larvae were removed from the damaged plants, the release of the low molecular weight compounds decreased immediately. In contrast, 16 hr later collection of volatiles from those damaged plants yielded large amounts of the indole, terpenes, and sesquiterpenes. Evidently, there is a delay in the plant's response to herbivore damage, which indicates changes in the plant's physiology and biochemistry. The actual mechanisms are still unknown, but it is obvious that rather than just passively emitting compounds from an open wound, the plant is actively releasing chemicals in response to herbivore feeding.

Artificially damaged plants do not emit large amounts of the terpenoids and indole emitted by the caterpillar damaged plants (Figure 1). However, application of caterpillar spit or regurgitant to the wound of

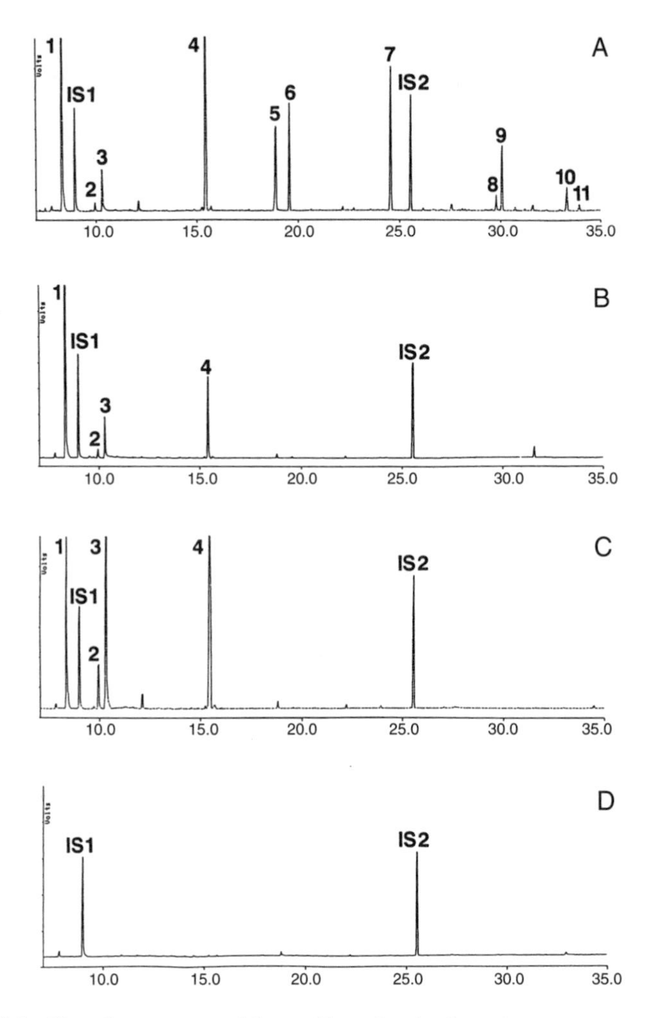

FIGURE 1 Gas chromatographic profiles of volatiles released by corn seedlings subjected to different damage treatments (from ref. 29). (*A*) Old damage: seedlings plus feeding beet armyworm caterpillars, where the caterpillars have been feeding on the seedlings overnight. (*B*) Fresh damage: seedlings plus feeding caterpillars (seedlings were undamaged at the start of volatile collection). (*C*) Artificially damaged fresh seedlings. (*D*) Undamaged seedlings. All collections lasted 2 hr. Peaks: 1, (*Z*)-3-hexenal; 2, (*E*)-2-hexenal; 3, (*Z*)-3-hexen-1-ol; 4, (*Z*)-3-hexen-1-yl acetate; 5, linalool; 6, (3*E*)-4,8-dimethyl-1,3,7-nonatriene; 7, indole; 8, α-*trans*-bergamotene; 9, (*E*)-β-farnesene; 10, (*E*)-Nerolidol; 11, (3*E*,7*E*)-4,8,12-trimethyl-1,3,7,11-tridecatetraene. Internal standards were *n*-octane (IS1) and *n*-nonyl acetate (IS2), each representing 1000 ng. Compounds were analyzed on a 50-meter methyl silicone capillary column. Ordinate is in volts; abscissa is in minutes of retention time.

artificially damaged leaves results in release of volatiles equal in quantity to those released by plants that have actually been fed on by caterpillars. Application of caterpillar spit to undamaged leaves does not induce volatile emission (30). Therefore, a substance in the caterpillar spit induces the wounded plants to begin making and releasing volatile chemicals. Moreover, choice tests in the wind tunnel revealed that artificially damaged plants with spit were as attractive to the wasps as the plants with real caterpillar damage. Without spit, the artificially damaged plants are much less attractive.

We also found that the response of the plant to the caterpillar spit is systemic (31). Thus, not only the damaged leaves but the entire plant produces and releases volatile compounds when one or more leaves are attacked by caterpillars. Dicke *et al.* (7) had earlier found a similar effect in that undamaged leaves of a spider mite-injured plant attracted predatory mites. This systemic effect could be very significant in terms of enabling the natural enemies to locate their victims. It makes the plant under attack stand out from its neighbors and act as a beacon to foraging natural enemies.

While a wasp may use the chemical cues released by damaged plants to find its cryptic herbivorous hosts, the chemical signals will vary considerably if its hosts are feeding on different plant species. In modern agriculture with large plantings of a single species, this may not be a problem. However for parasitoids like *Microplitis croceipes*, whose hosts feed on a wide variety of plant species, the different plants will send out completely different signals. For example, cotton, cowpea, or soybean each produce a unique blend of volatile chemicals when fed on by corn earworm caterpillars. Furthermore, the composition of the volatiles can even differ when the hosts feed on different parts of the same plant (12). The volatiles released from damaged flowers or flower buds may differ considerably from those released from damaged leaves of the same plant. When a parasitoid attacks caterpillars of several different species of moths, as does *C. marginiventris*, the picture gets even more complicated. The same plant may release two different blends when fed on by two different species of caterpillars. Thus, feeding fall armyworm larvae induce corn seedlings to produce a distinctly different blend than do beet armyworm larvae, but both are still suitable hosts for *C. marginiventris* (29).

Obviously plant volatiles induced by herbivore feeding will be quite variable among all the combinations and permutations of herbivore species, plant species, plant parts, and growth stages. Thus, the predator or parasitoid is faced with the formidable task of finding its prey or host in many different habitats with a great variety of odors. Under these conditions it is not surprising that the hunters rely heavily on learning (8,

12, 32, 33), and their response to plant volatile mixtures is reinforced by hunting success.

Two recent studies in our laboratories indicate that the rate of emission of insect herbivore-induced volatiles from corn and cotton varies over the course of the day (T.C.J.T., A. Manukian, R. R. Heath, and J.H.T., unpublished data; ref. 52). Release of the induced volatiles followed a diurnal cycle, with peak emission occurring during the photophase, which corresponds to the period during which parasitoids and predators normally forage. These studies and others that indicate variations in volatile emission with different growth stages and different parts of the plants indicate that considerable caution should be exerted when plant volatile emissions are studied. Particularly when plant–arthropod interactions are being investigated it is important to consider the growth stage and the part of the plant involved in the interactions as well as the time of the day that the interactions occur.

Whether the release of volatiles by the plant has evolved to attract the natural enemies of the herbivores or whether these carnivores only exploit a plant defense mechanism aimed at the herbivore is currently under discussion. It is possible that the increase in volatile emission by plants under attack by herbivores is connected to physiological changes in the plant, which increase the level of toxins and/or antifeedants in plant tissues and deter further herbivory. In corn, leaves damaged by lepidopterous larvae become less palatable to the feeding larvae. This parallels the increase in volatile release by larvae-damaged corn leaves (34). Also, cotton varieties that release the greatest quantities of volatiles when leaves are damaged by larvae are the least palatable to those larvae (J.H.L., A. Manukian, R. R. Heath, T.C.J.T., and J.H.T., unpublished data). If these phenomena are general, it may prove difficult to determine whether plants have evolved signals to attract natural enemies of the herbivores, or whether predators and parasitoids merely exploit a plant defense mechanism to find their herbivorous prey.

Price (35) predicted that future research will reveal widespread use of plant volatiles by herbivore predators, but it has only been looked for in a few systems. Mealybug infestations of cassava induce unidentified changes in infested and uninfested leaves which make both types of leaves more attractive to encyrtid wasp parasitoids (36). This system indicates that sucking as well as chewing insects can induce systemic changes in plant volatiles. Although theoretically less likely (8), some generalist predators appear to make some use of plant volatiles to locate prey. Yellow jacket wasps in the *Vespula vulgaris* species group are attracted to mixtures of (*E*)-2-hexenal with either α-terpineol or linalool—all ubiquitous components of volatiles from plants under attack by herbivores (37, 38).

DECEIT

The predators discussed up to this point search for prey by using their ability to perceive certain chemical clues. Some unusual predators have evolved the ability to attract their prey with scents that mimic the odor of a valuable resource (see reviews of chemical mimicry in refs. 9 and 39). Several groups of spiders lure male insect prey with scents that mimic the sex pheromone scents of females of the prey species (see reviews in refs. 9, 13, 40, and 41). To the best of our knowledge, these spiders are the only predators that mimic sex pheromones. However, the spiders share some similarities with the diverse orchids which mimic insect sex pheromones to lure pollinators (9, 42, 43) and with the predatory fireflies, which practice elaborate mimicry of visual sexual signals to lure their prey: heterospecific male fireflies (44).

While most spiders are generalist predators, the "bolas spiders" in the araneid subfamily Mastophorinae feed as adults almost exclusively on male moths (see reviews in refs. 9, 13, and 41). Hunting spiders construct a simplified web which includes a short dangling line ending in a drop of glue (the "bolas"), which is hurled at prey. Moths approach the spiders from down-wind, flying in a zig-zag, apparently anemotactic, fashion until the moth is within a short distance of the spider. Spiders can only capture moths that come within the range of the bolas (typically only three or four spider body lengths). Starting with the first published report of this behavior in 1903 (45), observers have suggested that the spiders mimicked the scent of female moths.

The research necessary to support this hypothesis has been hindered by the fact that the spiders are rarely encountered—they are nocturnal, cryptic, and appear to exist at very low population levels in most areas. These spiders are also difficult to work with in the field, where they move frequently, and in the laboratory, where they rarely engage in normal hunting behavior. Nonetheless, in recent years, field tests have shown that *Mastophora* species attract moths from five families (9, 13, 41, 46). At one field site in southern California where spiders survive the winter and hunting adults can be found throughout the year, *Mastophora cornigera* spiders caught at least 15 moth species (M.K.S. and W. Icenogle, unpublished data). The pheromone chemistry of eight of these moth species is known from chemical analysis of glands or volatiles or from field screening.

Identification of compounds in volatiles collected from hunting *M. cornigera* revealed three common components of moth sex pheromone blends: (*Z*)-9-tetradecenal, (*Z*)-9-tetradecenyl acetate, and (*Z*)-11-hexadecenal [while there was insufficient material for mass spectrometry, gas chromatographic retention time evidence suggests that (*Z*)-11-hexadece-

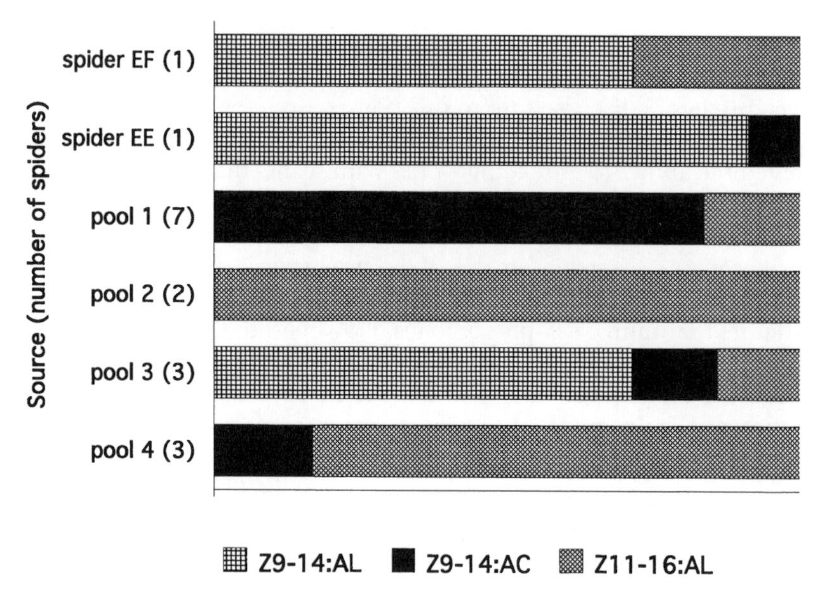

FIGURE 2 Relative proportions of three identified components of *M. cornigera* spider volatiles (47). The variation in the ratios suggests that spiders change the moth-attracting blends they produce. Z11-16:AL, (*Z*)-11-hexadecenal; Z9-14:AC, (*Z*)-9-tetradecenyl acetate; Z9-14:AL, (*Z*)-9-tetradecenal.

nyl acetate was also present] (47). These compounds are pheromone components for four *M. cornigera* prey species. In moths they are produced by the same biosynthetic pathway (48).

Volatiles collected from two single *M. cornigera* individuals and pools of collections from up to seven individuals contained quite different proportions of components (Figure 2). This preliminary evidence suggests that blends vary between individuals or that single individuals change blend composition over time. Field evidence also suggests that spiders vary their blends, since the spiders' prey includes many species that appear to be pheromonally incompatible—compounds that are necessary components in attractive blends for one or more moth species make blends unattractive for other species (Figure 3).

It is likely that the spectrum of compounds produced by all *Mastophora* species includes a wider range of compounds than those found in this study. Several *Mastophora* species catch male moths that are known to respond to pheromone compounds in the even-carbon number aldehyde/acetate/alcohol chemical class, as well as males of other species that are known to respond to pheromone compounds in the odd-carbon number hydrocarbon chemical class (9, 13).

Recent field tests have examined the prey-attracting ability of *Masto-*

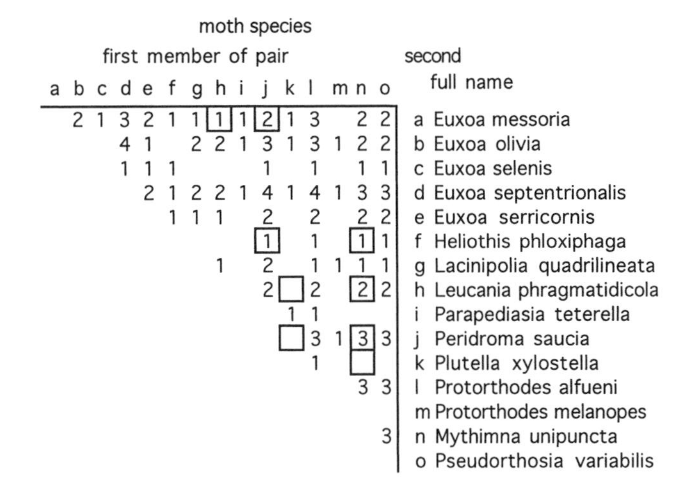

FIGURE 3 Pairwise combinations of moths caught by a population of spiders (*M. cornigera*) at one site in California (Winchester, CA) and the number of spider individuals catching both species in a pair [from a study of six spiders whose prey was recorded for up to 1.5 years over a span of 5 years (M.K.S. and W. Icenogle, unpublished data)]. Cells surrounded by a border indicate pairs of moths that seem to be pheromonally incompatible—i.e., compounds that are necessary components in attractive blends for one moth species make blends unattractive for the other species (47).

phora spiderlings as well as spiderlings and adults of two other spider genera (*Kaira* and *Phoroncidia*) that spin simplified webs (reviewed in refs. 9 and 13; see also ref. 49). These controlled tests demonstrate that in all three genera, adult spiders attract moths, and spiderlings attract various families of nematocerous flies. Preliminary chemical analysis suggests that adult *Phoroncidia* spiders produce known moth sex pheromone compounds (M.K.S. and J.H.T., unpublished data). These three spider genera are currently placed in unrelated subfamilies in the superfamily Araneoidea. Phylogenetic studies based on behavioral and morphological characters support this classification and the idea that the genera evolved the ability to attract prey independently (J. Coddington and N. Scharff, personal communication). (See references in ref. 9 for evidence of sex-pheromone mimicry in other groups of spiders.)

FUTURE DIRECTIONS

Many fundamental questions remain to be investigated in the study of these systems. How do reception and perception of pheromone molecules by "eavesdroppers" compare to the same processes in the prey? For example, male moths generally respond to only a range of pheromone component ratios that centers around the ratio of components produced by females (reviewed in ref. 50); do the wasps that eavesdrop on female moth sex pheromones have the same ratio specificity? How does this differ between specialist and generalist parasitoids? Does learning modify the responses of eavesdropping parasitoids and predators as it does in the parasitoids and predators that respond to plant "alarm" odors? In plants damaged by herbivorous arthropods, what are the mechanisms involved in the release of volatile signals? (See refs. 51, 52.) What are the factors in insect spit that plants respond to? How do hormones or other chemical signals bring about the systemic response of the entire plant to localized herbivore damage? How does the mechanism of plant response vary between major plant groups? (For speculation concerning the evolution of the eavesdropping and alarm systems, see refs. 8 and 10.) In the prey-attracting spiders, where are the volatile-producing glands? How does the biosynthesis of spider compounds compare to the biosynthesis of pheromone compounds by their prey? Do all three spider groups vary their prey-attracting blends? Is this variation controlled by learning, influenced by seasonal cues, or due to genetic differences between individuals? (For speculation concerning the evolution of these systems, see ref. 9.)

Much of the interest in these complex chemically mediated relationships lies in the potential agricultural application of research results. Future work on "eavesdropping"- and "alarm"-based relationships may lead to more effective use of parasitoids for biological control. Semiochemicals may be useful for attracting parasitoids or predators into a crop or increasing the amount of time they spend searching for hosts or prey in a field. New crop varieties may be developed that emit greater amounts of "alarm" compounds when attacked by herbivorous pests and thus are more effective in recruiting natural enemies of their attackers. Spiders that practice deceit represent an unexploited chemical library of compounds that may be useful in pheromone-based monitoring and control of crop pests. Analysis of spider volatiles might provide the first insight into the sex pheromone chemistry of those numerous prey taxa whose chemistry has never been studied or might reveal the existence of attractive sex pheromone analogs. The eventual applications arising from research in this area may be as unanticipated and exciting as many of the research results obtained so far.

SUMMARY

Arthropods that prey on or parasitize other arthropods frequently employ those chemical cues that reliably indicate the presence of their prey or hosts. Eavesdropping on the sex pheromone signals emitted to attract mates allows many predators and parasitoids to find and attack adult insects. The sex pheromones are also useful signals for egg parasitoids since eggs are frequently deposited on nearby plants soon after mating. When the larval stages of insects or other arthropods are the targets, a different foraging strategy is employed. The larvae are often chemically inconspicuous, but when they feed on plants the injured plants respond by producing and releasing defensive chemicals. These plant chemicals may also serve as "alarm signals" that are exploited by predators and parasitoids to locate their victims. There is considerable evidence that the volatile "alarm signals" are induced by interactions of substances from the herbivore with the damaged plant tissue. A very different strategy is employed by several groups of spiders that remain stationary and send out chemical signals that attract prey. Some of these spiders prey exclusively on male moths. They attract the males by emitting chemicals identical to the sex pheromones emitted by female moths. These few examples indicate the diversity of foraging strategies of arthropod predators and parasitoids. It is likely that many other interesting chemically mediated interactions between arthropod hunters and their victims remain to be discovered. Increased understanding of these systems will enable us to capitalize on natural interactions to develop more ecologically sound, environmentally safe methods for biological control of insect pests of agriculture.

REFERENCES

1. Law, J. H. & Regnier, F. E. (1971) *Annu. Rev. Biochem.* **40,** 533–548.
2. Nordlund, D. A. (1981) in *Semiochemicals: Their Role in Pest Control*, eds. Nordlund, D. A., Jones, R. L. & Lewis, W. J. (Wiley, New York), pp. 13–23.
3. Vinson, S. B. (1984) in *Chemical Ecology of Insects*, eds. Bell, W. J. & Cardé, R. T. (Sinauer, Sunderland, MA), pp. 111–124.
4. Noldus, L. P. J. J. (1989) Ph.D. dissertation (Wageningen Agricultural Univ., Wageningen, The Netherlands).
5. Nordlund, D. A., Lewis, W. J. & Altieri, M. A. (1988) in *Novel Aspects of Insect-Plant Interactions*, eds. Barbosa, P. & Letourneau, D. K. (Wiley, New York), pp. 65–90.
6. Turlings, T. C. J., Tumlinson, J. H. & Lewis, W. J. (1990) *Science* **250,** 1251–1253.
7. Dicke, M., Sabelis, M. W., Takabayashi, J., Bruin, J. & Posthumus, M. A. (1990) *J. Chem. Ecol.* **16,** 3091–3118.
8. Vet, L. E. M. & Dicke, M. (1992) *Annu. Rev. Entomol.* **37,** 141–172.
9. Stowe, M. K. (1988) in *Chemical Mediation of Coevolution*, ed. Spencer, K. (Academic, San Diego), pp. 513–580.
10. Dicke, M. & Sabelis, M. W. (1993) in *Insect Chemical Ecology: An Evolutionary Approach*, eds. Roitberg, K. & Isman, M. B. (Chapman & Hall, New York), pp. 122–155.
11. Tumlinson, J. H., Turlings, T. C. J. & Lewis, W. J. (1992) *Agric. Zool. Rev.* **5,** 221–252.

12. Turlings, T. C. J., Wäckers, F. L., Vet, L. E. M., Lewis, W. J. & Tumlinson, J. H. (1993) in *Insect Learning: Ecological and Evolutionary Perspectives*, eds. Lewis, A. C. & Papaj, D. R. (Chapman & Hall, New York), pp. 51–78.
13. Yeargan, K. V. (1994) *Annu. Rev. Entomol.* **39,** 81–99.
14. Wood, D. L. (1982) *Annu. Rev. Entomol.* **27,** 411–446.
15. Kennedy, B. H. (1984) *J. Chem. Ecol.* **10,** 373–385.
16. Klun, J. K., Plimmer, J. R., Bierl-Leonhardt, B. A., Sparks, A. N., Primiani, M., Chapman, O. L., Lee, G. H. & Lepone, G. (1980) *J. Chem. Ecol.* **6,** 165–175.
17. Lewis, W. J., Nordlund, D. A., Gueldner, R. C., Teal, P. E. A. & Tumlinson, J. H. (1982) *J. Chem. Ecol.* **8,** 1323–1331.
18. Noldus, L. P. J. J. & van Lenteren, J. C. (1985) *J. Chem. Ecol.* **11,** 781–791.
19. Noldus, L. P. J. J., Potting, R. P. J. & Barendregt, H. E. (1991) *Physiol. Entomol.* **16,** 329–344.
20. Nordlund, D. A., Lewis, W. J. & Gueldner, R. C. (1983) *J. Chem. Ecol.* **9,** 695–701.
21. Aldrich, J. R. (1995) in *Chemical Ecology of Insects II*, eds. Cardé, R. T. & Bell, W. J. (Chapman & Hall, New York), pp. 318–363.
22. Aldrich, J. R. (1985) in *Semiochemistry: Flavors and Pheromones*, eds. Acree, T. E. & Soderlund, D. M. (de Gruyter, Berlin), pp. 95–119.
23. Orr, D. B., Russin, J. S. & Boethel, D. J. (1986) *Can. Entomol.* **118,** 1063–1072.
24. Dicke, M. & Sabelis, M. W. (1988) *Neth. J. Zool.* **38,** 148–165.
25. Takabayashi, J., Dicke, M. & Posthumus, M. A. (1991) *Chemoecology* **2,** 1–6.
26. Takabayashi, J., Dicke, M. & Posthumus, M. A. (1991) *Phytochemistry* **30,** 1459–1462.
27. Turlings, T. C. J., Tumlinson, J. H., Eller, F. J. & Lewis, W. J. (1991) *Entomol. Exp. Appl.* **58,** 75–82.
28. Turlings, T. C. J., Tumlinson, J. H., Heath, R. H., Proveaux, A. T. & Doolittle, R. E. (1991) *J. Chem. Ecol.* **17,** 2235–2251.
29. Turlings, T. C. J. (1990) Ph.D. dissertation (Univ. of Florida, Gainesville).
30. Turlings, T. C. J., McCall, P. J., Alborn, A. T. & Tumlinson, J. H. (1993) *J. Chem. Ecol.* **19,** 411–425.
31. Turlings, T. C. J. & Tumlinson, J. H. (1992) *Proc. Natl. Acad. Sci. USA* **89,** 8399–8402.
32. Lewis, W. J. & Tumlinson, J. H. (1988) *Nature (London)* **331,** 257–259.
33. Lewis, W. J., Vet, L. E. M., Tumlinson, J. H., van Lenteren, J. C. & Papaj, D. R. (1990) *Environ. Entomol.* **19,** 1183–1193.
34. Turlings, T. C. J. & Tumlinson, J. H. (1991) *Fla. Entomol.* **74,** 42–50.
35. Price, P. W. (1981) in *Semiochemicals: Their Role in Pest Control*, eds. Norlund, D. A., Jones, R. L. & Lewis, W. J. (Wiley, New York), pp. 251–279.
36. Nadel, H. & van Alphen, J. J. M. (1987) *Entomol. Exp. Appl.* **45,** 181–186.
37. Aldrich, J. R., Kochansky, J. P. & Sexton, J. D. (1985) *Experientia* **41,** 420–422.
38. Aldrich, J. R., Lusby, W. R. & Kochansky, J. P. (1986) *Experientia* **42,** 583–585.
39. Dettner, K. & Liepert, C. (1994) *Annu. Rev. Entomol.* **39,** 129–154.
40. Eberhard, W. G. (1980) *Psyche* **87,** 143–169.
41. Stowe, M. K. (1986) in *Spiders Webs: Behavior and Evolution*, ed. Shear, W. (Stanford Univ. Press, Stanford, CA), pp. 101–131.
42. Kullenberg, B. (1961) *Zool. Bidr. Uppsala* **34,** 1–330.
43. Borg-Karlson, A. K. (1990) *Phytochemistry* **29,** 1359–1387.
44. Lloyd, J. E. (1984) in *Oxford Survey of Evolutionary Biology*, eds. Dawkins, R. & Ridley, M. (Oxford Univ. Press, Oxford), pp. 48–84.
45. Hutchinson, C. (1903) *Sci. Am.* **10,** 172.
46. Yeargan, K. V. (1988) *Oecologia* **74,** 524–530.

47. Stowe, M. K., Tumlinson, J. H. & Heath, R. R. (1987) *Science* **236,** 964–967.
48. Bjostad, L. B. & Roelofs, W. L. (1983) *Science* **220,** 1387–1389.
49. Eberhard, W. G. (1981) *J. Arachnol.* **9,** 229–232.
50. Phelan, P. L. (1993) in *Insect Chemical Ecology: An Evolutionary Approach,* eds. Roitberg, K. & Isman, M. B. (Chapman & Hall, New York), pp. 265–314.
51. Croft, K. P. C., Jüttner, F. & Slusarenko, A. J. (1993) *Plant Physiol.* **101,** 13–24.
52. Loughrin, J. H., Manukian, A. Heath, R. R., Turlings, T. C. J. & Tumlinson, J. H. (1994) *Proc. Natl. Acad. Sci. USA* **91,** 11836–11840.

Polydnavirus-Facilitated Endoparasite Protection Against Host Immune Defenses

MAX D. SUMMERS AND SULAYMAN D. DIB-HAJJ

Polydnaviruses not only have a unique taxonomic classification because of their segmented double-stranded DNA genomes but also exhibit an unusual relationship to two insects, an endoparasitic wasp and its host (1, 2). The virus is apparently symbiotically associated with the wasp where it is nonrandomly integrated into the chromosomal DNA of each male and female wasp in the wasp population. The virus replicates in specialized calyx cells of the female wasp at a specific time during late pupal development and in adults (3–7). After replication and assembly in the nucleus of the calyx cells, the polydnavirus is "secreted" into the wasp oviduct where it accumulates along with the wasp egg and a complex of oviduct secretions. Upon oviposition, the virus is transferred to a permissive host in which the endoparasite egg will develop. (Permissiveness in this context refers to a host that supports the development of the endoparasite.)

Significant developmental and physiological changes are induced in the parasitized host insect: phenomena that have been generally described as evidence of wasp-induced "host regulation" or "parasite-directed host manipulation" (8–10). The biological and biochemical

Max Summers is distinguished professor of entomology and director of the Center for Advanced Invertebrate Molecular Sciences at Texas A&M University, College Station. Sulayman Dib-Hajj is associate research scientist in the Department of Neurology, Neuroscience & Regeneration Research Center at Yale University School of Medicine, New Haven Connecticut.

manifestations of parasite-induced or -directed host manipulation are quite complex and vary significantly according to the specific parasitic wasp and host insect species; these changes are referred to as "immunosuppression" and "developmental arrest" (11, 12).

IMMUNOSUPPRESSION AND DEVELOPMENTAL ARREST

The endoparasite's strategy for survival involves avoidance of host recognition as foreign or inhibition of the ensuing immune reaction (12–17). Since encapsulation is apparently the major host defense against parasite egg invasion, host immunosuppression is usually attributed to the wasp's ability to avoid or suppress encapsulation. For the endoparasite *Campoletis sonorensis* Edson *et al.* (18) showed that purified viable polydnavirus was responsible for suppressing the host insect's (*Heliothis virescens*) ability to encapsulate the wasp egg. Theilmann and Summers (19) also showed that both immunosuppression and developmental arrest are induced in *H. virescens* by purified polydnavirus. Inhibition of encapsulation also correlated with a decrease in circulating hemocytes and reduced plasmatocyte attachment and "spreading" (20).

Upon recognizing a foreign object, hemocytes form a cellular sheath (capsule) around that object and neutralize it (2). Although the identity of all hemocyte classes involved in the encapsulation process is not clear, most of the attention in investigating polydnavirus-induced immunosuppression has been given to the role of granulocytes and plasmatocytes (2). Granulocytes and plasmatocytes are the two principal hemocyte classes known to participate in forming the capsule (14). Granulocytes are thought to be the first hemocyte cell type to reach the foreign object (13, 14). Upon recognizing the invading object—for example, an endoparasite egg—granulocytes degranulate, releasing chemotactic factors that attract more granulocytes and plasmatocytes. Plasmatocytes attach and spread on the surface of the foreign body, for example an endoparasite egg. The role of the host prophenol oxidase cascade and/or other humoral factors in recognition of the endoparasite egg and initiation of encapsulation is not well defined (13).

DIVERSE STRATEGIES FOR SUCCESSFUL PARASITIZATION

In the general spectrum of host–parasite relationships, there is considerable parasite-induced host variability resulting in perturbed host physiology, biochemistry, and developmental behavior (8, 21, 22). There is a wide range of speculation that the parasite has the capacity to modulate or regulate host systems (8, 10, 23). It is clear that the *C. sonorensis* polydnavirus induces developmental arrest in *H. virescens* larva (18, 24,

25), which is manifest as a significant reduction in weight gain and the delay of pupation to extend host larval life for a period of 9–12 days (19, 26). The mechanisms employed by polydnaviruses to alter host development have been comprehensively documented (8, 9) and must be studied pursuant to the identification of viral-induced or viral-expressed products that function in that role.

Parasitic wasps introduce several oviduct and venom gland secretions along with the wasp egg into the host. There are endoparasitic Hymenoptera that apparently alter the host's immune and developmental processes by parasite venom gland secretions (11, 27–30) or by both venom secretions and polydnaviruses (31–33). Edson *et al.* (18) showed that successful parasitization of *H. virescens* is dependent on *C. sonorensis* polydnavirus gene expression. Endoparasite eggs that are washed in physiological saline and injected into *H. virescens* alone or in combination with UV-inactivated *C. sonorensis* polydnavirus are encapsulated and killed. Preincubation with monoclonal antibodies specific for viral envelope proteins neutralized the effects of the virus on *H. virescens* (34). However, injecting *C. sonorensis* eggs with either the calyx fluid from the wasp oviduct or gradient-purified *C. sonorensis* polydnavirus results in the successful development of the endoparasite. So, *C. sonorensis* polydnavirus gene expression is both necessary and sufficient for inducing *H. virescens* immunosuppression and developmental arrest and, therefore, is essential for parasite survival.

POLYDNAVIRUSES: GENERAL FEATURES

The major characteristic that distinguishes polydnaviruses from other animal DNA viruses is a segmented genome of multiple, covalently closed circular, double-stranded DNAs (6, 35, 36). Polydnaviruses have only been isolated from species of endoparasitic wasps belonging to the hymenopteran families of Braconidae and Ichneumonidae. The DNA genomes of viruses from different species of wasp endoparasites differ in the number of DNA segments and their molar abundance and sizes. A comparison of different polydnavirus species reveals from <10 to >25 segments in the genome (2, 6, 37). It is difficult to determine polydnavirus genetic complexity because not only do segments vary in number and size but also related sequences are found on the same segment and on different segments. The best estimates of the genome sizes are from 75 kbp to >250 kbp (6). The structure of the genome and nature of the virus life cycle clearly show that it is genetically very complex.

The nucleocapsid of the *C. sonorensis* polydnavirus is prolate ellipsoid in shape and has two envelopes (37): one envelope is obtained in the nucleus and the other as the virus buds through the calyx cell membrane into the lumen of the oviduct. The genome consists of ≈28 segments. Each DNA segment is given an alphabetical designation from A to W in order of increasing size. The segments are not present in equimolar ratios, but the agarose gel profiles of the banding patterns are qualitatively and quantitatively constant. Despite the genomic complexity, several rare and abundant genomic segments were cloned in their entirety, which allowed the mapping and sequencing of individual viral genes to be described in this paper (4, 25, 38). The functional significance of the different molar abundance of the viral segments is not clear at this time, but it may reflect the mode of replication and/or recombination among the multiple segments.

Viral DNA is integrated into the chromosomal DNA of male and female wasps. The virus also exists in an episomal form in female tissues and even in male wasps, though to a much reduced level (4). It is not known whether the linear or the episomal forms of the virus are the templates for viral replication.

POLYDNAVIRUS EXPRESSION IN THE HOST

Although there is no detectable replication of the polydnavirus in *H. virescens*, viral DNA persists in infected tissues (19, 39), and there is abundant expression of several viral mRNAs. Fleming *et al.* (40) provided the first evidence that polydnavirus genes are transcribed in parasitized *H. virescens* larvae. At least 12 size classes of viral mRNAs are detected by Northern analysis in parasitized *H. virescens* during the course of endoparasite development (9–12 days), with some viral mRNAs detected as early as 2 hr postparasitization (hpp) (25, 38, 41–44). This led to the genomic mapping of the most abundant viral mRNAs to polydnavirus segment W (38). The mapping and sequencing of a number of viral genes and computer-assisted analysis of nucleotide and predicted amino acid sequences provided several additional insights regarding the function of the viral expression products. At this point, the most significant of these observations can be summarized as follows: (*i*) Transcription in the segmented genome is multipartite in that a specific cDNA probe will hybridize to different genomic segments. Comparisons of the DNA sequences of these cross-hybridizing genes revealed that there are related yet different genes on the same genomic segment and, very likely, on other segments. (*ii*) Different cDNA probes hybridized to different but sometimes partially overlapping sets of viral DNA segments. (*iii*) Almost all of the *C. sonorensis* polydnavirus genes mapped and studied to date belong to gene families. (*iv*) The expression of these genes apparently occurs in a host-specific manner, with some genes expressed only in the parasitized host and some only in the wasp, whereas others are expressed in both insects.

The purpose of our studies is to identify the viral expression products responsible for parasite survival of the host immune response. To do this, we have focused on identifying those viral genes that are abundantly expressed early (2 hpp) and continue during parasitization. We have been successful in mapping, cloning, and sequencing several of these genes and their corresponding cDNAs. We have cloned and expressed these cDNAs in the baculovirus expression system, and we have now formulated a working hypothesis to explain the role of a subset of these gene products.

VIRAL MULTIGENE FAMILIES

So far, most of the *C. sonorensis* polydnavirus genes that are expressed in *H. virescens* are grouped into two gene families: "repeat" (25) and "cysteine-rich" (45) gene families. An additional "venom-related" gene family was identified based on immunological relatedness (34). The organization of several genes into families in *C. sonorensis* polydnavirus is yet to be described in other polydnaviruses. Although genes of gene families exist in other animal viral genomes (46–49), it is unusual to find different multigene families in the same virus genome.

CYSTEINE-RICH GENE FAMILY

The cysteine-rich gene family is presently defined by the viral genes W*Hv*1.0, W*Hv*1.6, and V*Hv*1.1 [nomenclature: segment, W or V; host, *Hv* (*H. virescens*); RNA size in kb] (45). These genes share a common gene structure including introns at comparable positions and encode secreted proteins that contain one or two cysteine motifs of the pattern C-C-CC-C-C (38, 45). W*Hv*1.0 and W*Hv*1.6 each contain a single cysteine motif, and V*Hv*1.1 contains two motifs. The sequence encoding the cysteine motif is interrupted by an intron immediately preceding the codon of the second cysteine of the motif. Also, introns interrupt the 5′ untranslated leader at comparable positions. The conservation of the intron positions clearly indicates that the three viral genes have a common ancestry. W*Hv*1.0 and W*Hv*1.6 are more related to each other than either is related to V*Hv*1.1; therefore, we grouped these genes into two subfamilies. A very unusual feature in the W*Hv*1.0/W*Hv*1.6 subfamily is that their intron 2 sequences are more conserved, over their entire length, than their flanking exons (45).

GENE STRUCTURE OF W*Hv*1.0, W*Hv*1.6, AND V*Hv1.1*

The cysteine-rich gene family was initially discovered as two separate gene families (41). Transcripts of 1.0 and 1.6 kb (encoded by W*Hv*1.0 and

W*Hv*1.6, respectively) and 1.1 and 1.4 kb (encoded by V*Hv*1.1 and a gene yet to be identified, respectively) from parasitized *H. virescens* hybridized, under high-stringency conditions, to two different *C. sonorensis* polydnavirus genomic clones. Multiple viral genomic segments hybridized to these genomic clones. Those results suggested that two different and potentially large gene families are encoded by a discrete set of viral genomic segments. The reduced homology of W*Hv*1.0 and W*Hv*1.6 to V*Hv*1.1 sequences (45) explains the lack of cross-hybridization to the 1.1- and 1.4-kb transcripts under high-stringency conditions (38, 41). Additional members of this subfamily probably exist on other genomic segments as indicated by Southern analysis (38, 41).

The relatedness of W*Hv*1.0 and W*Hv*1.6 was confirmed when the cDNAs for these genes were cloned and sequenced (38). Five regions (A–E) were found to be 68–80% identical at the nucleotide and amino acid levels. These conserved regions were separated by regions of very low similarity. Region A encompasses the 5′ untranslated leader and the first 16 amino acids of the predicted open reading frame. Region B encompasses a stretch of 27 amino acids immediately preceding the cysteine motif and the primary sequence delineated by the first three cysteine residues of this motif. Region C encompasses the sequence delineated by the last two cysteine residues of this motif and the C terminus. Regions D and E are located in the 3′ untranslated sequence, separated by over 400 bp. After we cloned and sequenced V*Hv*1.1 and identified the conserved cysteine motif and locations of introns in all three genes, we reevaluated and extended the data for W*Hv*1.0/W*Hv*1.6 (45).

The promoter, 5′ and 3′ untranslated sequences, and introns 1 and 2 of W*Hv*1.0 and W*Hv*1.6 share 80% or greater similarity (identity) (38, 45). This may be explained by a relatively recent intrasegment gene duplication event. By contrast, the sequences encoding the mature N and C termini of these two proteins have significantly diverged (44% and 52% similarity, respectively) (38). The molecular mechanisms that maintain conserved noncoding sequences, including introns, but allow significant divergence of exon sequences are not clear at this time. The length difference between the 1.0- and 1.6-kb transcripts is due to 577 nt in the 3′ untranslated sequence, which may have resulted from two insertions/deletions (38).

A distinctive feature of this gene family is the high degree of similarity (92%) in intron 2 sequences of W*Hv*1.0 and W*Hv*1.6. This similarity is higher than that (76%) of the immediately flanking exon sequences, which encode the cysteine motif (45). The nucleotide similarity spans the whole intron sequence with two insertions/deletions toward the 5′ and 3′ ends of the introns, which are required to optimize the sequence alignment. A high degree of sequence similarity among introns of recently duplicated genes was reported, but in those cases the exon sequences were conserved to the same degree (50, 51).

It is very unusual for introns to be more conserved than their flanking exons. Except for the conserved splicing signals at the 5′ and 3′ ends, and the branch site of the intron, the primary intron sequence is not thought to play an important role in splicing (52). Also, unlike the cases of other nuclear pre-mRNA introns, computer-assisted analysis did not reveal the presence of known control elements (53–58) or the presence of functional transcripts or independent genes (59–61). The high degree of sequence conservation may simply reflect the short time since the divergence of the two genes. Alternatively, these introns may have a functional role yet to be identified.

The conservation of the intron position with respect to the cysteine motifs of V*Hv*1.1 strongly indicates an intragenic duplication event. Introns 2 and 3 of V*Hv*1.1 show similarity to intron 2 in W*Hv*1.0 and W*Hv*1.6, which is limited to the 5′ and 3′ ends beyond the conserved splicing signals: these introns also show the same limited similarity to each other (45). Assuming that introns in these genes are subject to similar evolutionary forces, the limited similarity between introns 2 and 3 of V*Hv*1.1 indicates that the presumed intragenic duplication preceded the duplication of W*Hv*1.0 and W*Hv*1.6.

Alternative splicing is another control process involved in gene expression that can produce different proteins from the same gene (62, 63). We isolated a V*Hv*1.1-like cDNA, pcVR900 (34), which utilizes an alternative 3′ acceptor site of intron 1 and an alternative polyadenylylation site. That cDNA, however, is incomplete; it lacks an initiator methionine at an analogous position to V*Hv*1.1, and the 5′ hexanucleotide consists of thymidine residues not coded by the V*Hv*1.1 gene (S.D.D.-H., B. Webb, and M.D.S., unpublished results). By using an internal AUG codon, that cDNA has a predicted open reading frame of only 60 amino acids that bear no similarity to the 217 amino acids of V*Hv*1.1. It is possible that this cDNA may have resulted from a cloning artifact, but it is also possible that the corresponding transcript may be encoded by another V*Hv*1.1-like gene or may be produced by trans-splicing of a leader sequence. Alternative and trans-splicing are two mechanisms to generate genetic diversity that may be utilized by this virus.

OPEN READING FRAMES OF THE CYSTEINE-RICH GENES

The cysteine motifs in this polydnavirus gene family are structurally analogous to the cysteine motifs of the ω-conotoxins. The ω-conotoxins are high-affinity ligands with different receptor specificity for voltage-sensitive ion channels (64, 65). The mature ω-conotoxins have three disulfide bridges that function as a conserved, highly compact, structural scaffold with hypervariable intercysteine amino acid residues. The *C. sonorensis* polydnavirus cysteine-rich gene family encodes secreted proteins that are apparently produced throughout the parasitization period. The hypervariability in amino acids of the cysteine motif of these genes suggests that these proteins may bind to related, yet different, targets.

The sequences encoding the cysteine motifs in W*Hv*1.0 and W*Hv*1.6 are 76% identical, counting a 33-bp deletion as a single mutational event; however, the two motifs are 58% identical at the amino acid level (45). The relative high similarity at the nucleotide sequence level and the decreased similarity at the amino acid sequence level are inconsistent with predictions of the neutral theory of mutation (66). The majority of the nucleotide substitutions in this domain are in codon positions 1 and 2 and multicodon positions, which cause amino acid replacement. A similar pattern of nucleotide substitutions is found in the sequence encoding the antigen recognition sites of the major histocompatibility complex I and II genes (67, 68). Codon substitutions in the antigen recognition site sequences result in hypervariability that is biologically significant and is explained by overdominant selection or positive Darwinian selective pressure.

Nei and coworkers (69) developed a mathematical model to describe overdominant selection. Briefly, they determined the fraction of silent substitutions (synonymous changes per synonymous site, d_S) and then determined the fraction of replacement substitutions (nonsynonymous changes per nonsynonymous site, d_N). In the case of the antigen recognition sites, d_S is 2- to 4-fold smaller than d_N; however, d_S is significantly greater than d_N (5- to 9-fold higher) in the constant domains (67, 68). When the sequence encoding the cysteine motifs of W*Hv*1.0 and W*Hv*1.6 are compared to each other, d_S is only about 2-fold greater than d_N. But, when the codons invariant in these two motifs are excluded from the analysis, d_S and d_N become comparable (45). This analysis included the codons of all the motif amino acids because those involved in binding to the putative target are yet to be identified. Using all of the codons in the analysis may have masked overdominant selection, since amino acids not involved in binding to the putative target would not be subject to overdominant selection. Alternatively, no negative selective pressure is exerted on this motif. It is conceivable that both of these mechanisms are at work on different codons of the cysteine motif.

Aside from the cysteine motif, conserved regions in the proteins encoded by W*Hv*1.0 and W*Hv*1.6 are likely to reflect functional roles. Based on the rules of Von Heijne (70), the N-terminal 16 amino acids may function as a signal peptide for secretion. Recombinant W*Hv*1.0 and W*Hv*1.6 proteins are secreted into the medium of infected insect cells (43) and into the hemolymph of parasitized *H. virescens* (S.D.D.-H., B. Graham, and M.D.S., unpublished results). By analogy to the propeptides of conotoxins, the highly conserved precysteine domain (26 out of 27 identical amino acids) may be important for folding the cysteine motif into a structure compatible with optimal biological function. Alternatively, this domain may serve another function yet to be determined.

The predicted mature N termini of W*Hv*1.0 and W*Hv*1.6 are only 16% identical, whereas the sequences C-terminal to the cysteine motif are about 36% identical (38). The untranslated sequences in these genes are more conserved than the sequences encoding these regions. It is possible that the specific role of the N and C termini is largely independent of their primary sequence. Alternatively, the divergence may reflect possible different functions. These conclusions will be further clarified as other members of this subfamily are cloned and sequenced or as functional assays for these proteins are developed. The fact that the N and C termini are flanked by highly conserved sequences may indicate that homologous recombination among members of this family results in the exchange of these domains to generate novel combinations.

In addition to the two cysteine motifs, the open reading frame of V*Hv*1.1 has three degenerate motifs, EPEADGKT, DEAN, and SAT (45) characteristic of the "DEAD" family of ATP-dependent RNA helicases (71, 72). The invariant amino acids of these motifs are underlined. The translation initiation factor elF-4A is the prototype of this family (71); however, other family members are involved in splicing (73). The order but not the spacing of these motifs is conserved in the V*Hv*1.1 open reading frame. The first and second V*Hv*1.1 motifs are analogous to the DEAD family "A" and "B" motifs, which are essential for ATP binding and hydrolysis (74–76). The third motif, SAT, is implicated in RNA unwinding (75). Whether these motifs are functional in V*Hv*1.1 remains to be seen. The presence of these motifs, separated by the N-terminal cysteine motif, may better be explained by convergent evolution than the acquisition of a host DEAD domain by the virus.

VENOM-RELATED GENE FAMILY

Webb and Summers (34) report that polydnavirus envelope proteins share cross-reacting epitopes with proteins from wasp venom glands. The conservation of epitopes on the virus and venom gland proteins may suggest functional or evolutionary relationships. Three isolates of monoclonal antibodies raised against venom gland proteins of *C. sonorensis* were selected because each cross-reacted with *C. sonorensis* polydnavirus envelope proteins and soluble wasp oviduct and venom gland proteins. The viral envelope proteins had different molecular weights compared to wasp venom and soluble oviduct proteins. Preincubation of purified *C. sonorensis* polydnavirus with these monoclonal antibodies neutralizes the effects of the virus when injected into *H. virescens*. The immunological relatedness of different viral envelope proteins to each other and to wasp venom proteins defines the venom-related gene family.

The viral genes encoding the envelope proteins and the genes for the venom secretions are yet to be isolated and studied. The existence of the venom-related gene family was also indicated by the isolation of cross-hybridizing cDNAs from a wasp and parasitized *H. virescens* libraries (34). However, that result may have been fortuitous because the presence of the sequence encoding the DEAD motifs in pcVR900 cDNA (B. Webb and M.D.S., unpublished results) may explain the cross-hybridization to the wasp cDNA. Sequencing of the wasp cDNA revealed the presence of all conserved features of elF-4A and a significant overall similarity to mouse and yeast cognates (B. Webb and M.D.S., unpublished results). The wasp cDNA is likely to correspond to the wasp elF-4A gene.

REPEAT GENE FAMILY

Early reports indicated that the *C. sonorensis* polydnavirus genome was made up of mostly unique sequences (35). However, Blissard *et al.* (41) demonstrated that different genomic segments hybridized to distinct subsets of viral segments. Further analysis indicated that sequence homology exists among the majority of genomic segments (44). Viral segment O^1 hybridized to 11 other DNA segments. Highly repetitive sequences believed to be largely, although not entirely, responsible for the extensive homology among the segments were identified on segments B, H, and O^1. An optimal sequence alignment revealed an imperfectly repeated consensus sequence $\approx$540 bp in length, with an average similarity of 60–70%. Shorter regions of sequences approach 90% similarity, and the two repeated sequences on segment O^1 share long stretches of sequence similarity. The 540-bp repeat elements are present as a single repeat, as on segment B, or in direct tandem arrays of two or more, as on segments H and O^1. Southern blot analyses of the viral genome under conditions of low stringency indicate that sequences related to the 540-bp element are conserved on most of the DNA segments.

Northern blot analyses indicate that sequences of the four repeated elements are present on transcripts expressed in *H. virescens* by 2 hpp (25). Steady-state RNA levels peak at 2–6 hpp and then decline over the next 8 days. These transcripts are not as abundant as the cysteine-rich transcripts but are still major viral expression products.

Sequence analysis of cDNA clones shows that the 540-bp consensus repeats on DNA segment B, H, and O^1 are contained within an open reading frame. These predicted open reading frames have been confirmed by *in vitro* translation and expression of recombinant proteins in bacteria. Putative signal peptides at the N terminus of the open reading frames are not predicted for any of the 540-bp repeat gene protein sequences.

These repeated elements share no significant homology to the cysteine-rich gene family. This family of genes also differs from the cysteine-rich gene family in that its members do not contain introns and three of the four transcripts studied so far are also expressed in oviducts of the female *C. sonorensis*. This, combined with the lack of significant similarity in the protein sequence, suggests that this family will likely play a role in the *C. sonorensis* polydnavirus life cycle distinct from that of the cysteine-rich gene family.

PARASITE–VIRUS–HOST

Many parasites or pathogens use insects as reservoirs, vectors, or hosts. Parasites also utilize diverse strategies to survive host defense mechanisms and must have the ability to rapidly evolve in response to them. A fundamental characteristic of strategies for parasite survival is the capability of rapid adaptation to their host immune defenses. The genetic basis for diversity within these strategies is the evolution of new genetic entities that involve gene amplification and regulation of gene expression at multiple levels, which may vary according to the specific parasite–host system (77–79). These include not only the expression of molecules designed for the passive protection of the parasite but also other effector molecules, which in turn can alter control and/or processing of gene expression products. Impose upon this the numerous parasite–host insect species and plethora of cells, tissues, and host developmental factors and processes that the parasite may target for its selective advantage, and one can appreciate the considerable potential for "diversity."

The parasite–host interaction continuously exerts selective pressure(s) on both insects to survive. The host insect presents a particularly challenging environment for endoparasites because of the rapid development and differentiation that characterize parasite and host life cycles (21, 22, 80, 81). Endoparasites may exploit a variety of agents to suppress or avoid host defenses and to modify the normal development of the host to match their need: polydnaviruses, venoms, oviduct secretions, and protective materials coating the parasite egg or produced by the endoparasite as it develops within its host.

In a permissive host, not all immature wasps are successful in surviving the host's defenses, indicating that selection pressures bearing on the survival strategies of both insects involve some balance that favors the parasite. To the extent that failed endoparasite development is not due to a defect in the wasp egg, it is likely that the surviving endoparasite represents a genetic lineage associated with a competent polydnavirus. For the polydnavirus of *C. sonorensis*, experimental data demonstrate that polydnavirus gene expression in the parasitized insect is sufficient for parasite survival. Assuming that the polydnavirus genetic system is in a dynamic state of evolution, one must be curious about the molecular mechanisms by which the surviving wasps are able to retain the essence of the most effective polydnavirus genetics to favor the wasp's survival strategy.

SEGMENTED VIRUS GENOMES: EVOLUTION AND FUNCTIONAL SIGNIFICANCE

The endoparasite *C. sonorensis* has evolved with the ability to generate extrachromosomal genetic elements in the form of multiple double-stranded, superhelical DNA molecules. These DNA molecules are amplified in the calyx cell nucleus, packaged into viruses, and secreted in a complex process of viral maturation, which also provides a complex double viral envelope. One viral envelope is assembled in the cell nucleus, and the other is obtained during budding from the calyx cell surface into the oviduct lumen. Viral envelopes, which are derived from cellular membranes, may mediate species-specific virus host cell and tissue interactions. This could be one important aspect of the species-specific endoparasite–host relationship fundamental to parasite survival.

Perhaps polydnaviruses are another variation upon diverse eukaryotic mechanisms of gene amplification and transfer of genetic information. Numerous examples of mechanisms of extrachromosomal gene amplification and maintenance of superhelical genomic DNA (82–96) and viral acquisition and adaptation of host genes to benefit the parasite or the virus (97–99) are reported in the literature. Also, portions of eukaryote genomes are believed to consist of sequences that correspond to transposable and mobile elements. Some of these are known to be widespread among insect orders (100). It is reasonable to consider that the heterodisperse genetic elements of a polydnavirus may represent mobile elements that have evolved to take a viral form during horizontal movement to another eukaryotic organism. Theoretically, this segmented complex of extrachromosomal DNA superhelices, and the genes encoded by them, could undergo rapid evolution (77, 93, 94, 101, 102).

It is also possible that polydnaviruses were originally virulent in *C. sonorensis*, but the virus–wasp relationship has evolved toward mutualism (103). Obligate symbiosis enables the host to acquire functions that improve its chances for survival; the mutualist also benefits by securing its passage to the host progeny. Polydnaviruses may be optimal mutualists in that their genomes are integrated into the wasp chromosomal DNA and, thus, transmitted vertically to each male and female in the wasp population (2, 6).

GENE FAMILIES: EVOLUTION AND SIGNIFICANCE

A significant part of the *C. sonorensis* polydnavirus segmented genome is apparently organized into several families of genes. It is not yet known if the occurrence of gene families is a general phenomenon for polydnaviruses, but *C. sonorensis* polydnavirus clearly possesses the genetic capability to express a large number of gene families, each with a significant amount of potential genetic variation. So, what is the significance of these viral gene families? It is recognized that gene families are a major source of genetic variability and that families evolve by gene duplication followed by sequence divergence (104). Gene duplication may occur during inter- and intramolecular recombination among the several viral genomic segments during viral DNA replication and/or by chromosomal crossing-over during wasp sexual reproduction. This viral genetic system presents significant potential for a large number of duplicated, yet diverging, genes encoding a variety of related, yet different, proteins.

Virus gene families have the potential to rapidly generate diversity to counter adaptive changes in host immune defenses to benefit the wasp's strategy for survival. We are limited in our ability to assess the significance of multigene families in the virus until the function(s) of the various proteins are identified. However, a classic example of the use of gene families is the surface antigen variability in trypanosomes (101, 102). In response to host selection pressure, trypanosomes express variable surface antigens by activating different members of a gene family.

Positive Darwinian selective pressure or overdominant selection is implicated in the evolution of the circumsporozoite protein genes of the malaria parasites, glutathione *S*-transferase of *Schistosoma* (105, 106) and the hemagglutinin genes of human influenza viruses (107). The vertebrate immune system has the potential to produce an impressive repertoire of molecules that recognize invading foreign antigens. The antigen-binding sites of antibodies and surface receptors encoded by the major histocompatibility complex I and II genes were also shown to be under positive Darwinian selective pressure (67, 68, 108). Nucleotide substitutions in the codons of the antigen-binding sites acquire more replacement than silent mutations. The ability of the immune system to generate this diversity places significant selective pressure on the parasite: parasite proteins are subjected to two opposing evolutionary forces that control amino acid substitution (101). The first force is for neutral selection favoring silent mutations that retain the function of that protein, particularly if it is a nonstructural protein. The second force is the pressure toward diversity to avoid recognition by the immune system.

We have applied these models to assess if overdominant selection underlies the hypervariability of the intercysteine amino acid residues of the cysteine motif of the *C. sonorensis* polydnavirus cysteine-rich gene family (45). Statistical tests applied to the sequence encoding these motifs in W*Hv*1.0 and W*Hv*1.6 are consistent with this possibility but are inconclusive. Our analyses, however, are limited by the lack of knowledge of the specific amino acids that interact with the target molecules. Once identified, more rigorous analyses are possible. In further support of a possible functional significance of the intercysteine amino acid hypervariability, the replacements in the intercysteine residues also change the overall charge distribution in this domain. Analysis of the cysteine motifs of the V*Hv*1.1 gene awaits cloning of other members of this subfamily.

The repeat gene family tandem repeats may increase both intra- and intermolecular recombination events, which result in gene or epitope amplification that is usually followed by sequence divergence. This could be another strategy of adaptation by the virus to keep pace with its evolving host. Because of its potentially large size and sequence diversity, the 540-bp repeat gene family will presumably play an important role. Based upon predicted properties, the 540-bp repeat proteins are believed to function differently than those of the cysteine-rich family.

CONCLUSIONS

Several *C. sonorensis* polydnavirus genes are expressed by 2 hpp and during parasite development. The abundance and temporal pattern of expression strongly suggest that these viral-encoded proteins provide important functions for successful parasitization. For that reason, we have investigated the identity and the molecular basis of the structure and function of these genes and their products. We have identified several members of the cysteine-rich and repeat gene families (25, 38, 41, 45). So far, it has been difficult to propose a role for the expression products of the repeat gene family. However, there is preliminary evidence that conserved sequence elements in a member of the repeat gene family on segment B may be involved in the integration/excision events of that segment (4). The hypervariability of the intercysteine amino acids of the polydnavirus cysteine-rich proteins and the analogy to the ω-conotoxins suggest that members of this gene family bind to different molecular targets in *H. virescens* to affect blood cell functions, the encapsulation process, or other host functions.

It is reasonable to propose that the cysteine-rich proteins may play an important role in preventing the recognition of foreign objects and/or the normal response of components of the immune system. Poxviruses encode soluble lymphokine receptors that are secreted from infected cells to act as decoys and prevent identification and killing of virus-infected cells (109). The cysteine-rich family of proteins may bind to analogous signals in the insect hemolymph and prevent the activation of the immune response. Alternatively, these soluble proteins may bind receptors on hemocyte surfaces and prevent recognition of the endoparasite as foreign or inhibit the normal response of those cells. In studies of parasite-induced immunosuppression of insects, a soluble "transformation" factor(s) is proposed to explain the altered behavior of plasmatocytes (20), which correlates with suppression of encapsulation. Our model is consistent with the presence of such factor(s). At later times when the host immune system appears to be significantly compromised, cysteine-rich proteins, through their hypervariable cysteine motif, may bind to related yet distinct targets to affect multiple host systems.

This paper is an assessment of our current knowledge of the structure and function of gene families of polydnavirus of *C. sonorensis* and their expression relative to their role in endoparasite survival. It is not intended to provide a comprehensive review of endoparasite–host relationships or the several polydnaviruses associated with them: the literature citations in the first part of this paper are quite sufficient in that regard. Only a few other polydnavirus–endoparasite–host systems are under experimental scrutiny. Polydnavirus-induced host alterations are being studied in *Manduca sexta* (8), *Trichoplusia ni* (110), *Spodoptera frugiperda* (111), *H. virescens* (112), and *Pseudoplusia includens* (39). However, the structural and functional organization of the polydnavirus genome of *C. sonorensis* is the most comprehensively studied at the molecular level. From these studies, we have sufficient information on gene structure and expression of several families of gene products to begin a more direct inquiry into their function. The possible involvement of the cysteine-rich gene proteins in targeting parasitized host systems to facilitate parasite survival is currently a testable model for the role of polydnavirus proteins in inducing host immunosuppression and developmental arrest. If successful, our knowledge of these functions may provide additional insights for the biochemistry and cell biology of insect cellular immune processes.

SUMMARY

The polydnavirus of *Campoletis sonorensis* has evolved with an unusual life cycle in which the virus exists as an obligate symbiont with the parasite insect and causes significant physiological and developmental alterations in the parasite's host. The segmented polydnavirus genome consists of double-stranded superhelical molecules; each segment is apparently integrated into the chromosomal DNA of each male and female wasp. The virus replicates in the nucleus of calyx cells and is secreted into the oviduct. When the virus is transferred to the host insect during oviposition, gene expression induces host immunosuppression and developmental arrest, which ensures successful development of the immature endoparasite. In the host, polydnavirus expression is detected by 2 hr and during endoparasite development. Most of the abundantly expressed viral genes expressed very early after parasitization belong to multigene families. Among these families, the "cysteine-rich" gene family is the most studied, and it may be important in inducing host manifestations resulting in parasite survival. This gene family is characterized by a similar gene structure with introns at comparable positions within the 5′ untranslated sequence and just 5′ to a specific cysteine codon (*C) within a cysteine motif, C-*C-CC-C-C. Another unusual feature is that the nucleotide sequences of introns 2 in the subfamily W*Hv*1.0/W*Hv*1.6 are more conserved than those of the flanking exons. The structures of these viral genes and possible functions for their encoded protein are considered within the context of the endoparasite and virus strategy for genetic adaptation and successful parasitization.

We thank Brent Graham, Sharon Braunagel, and Shelley Bennett for their critical evaluation of this paper. This study was funded in part by National Science Foundation Grant IBN-9119827 and Texas Agricultural Experiment Station Project 8078.

REFERENCES

1. Stoltz, D. B. & Vinson, S. B. (1979) *Adv. Virus Res.* **24,** 125–171.
2. Stoltz, D. B. (1993) in *Parasites and Pathogens of Insects*, eds. Beckage, N. E., Thompson, S. M. & Federici, B. A. (Academic, San Diego) Vol. 1, pp. 167–187.
3. Fleming, J. G. W. & Summers, M. D. (1986) *J. Virol.* **57,** 552–562.
4. Fleming, J. G. W. & Summers, M. D. (1991) *Proc. Natl. Acad. Sci. USA* **88,** 9770–9774.
5. Fleming, J. G. W. (1991) *Biol. Control* **1,** 127–135.
6. Fleming, J. G. W. & Krell, P. J. (1993) in *Parasites and Pathogens of Insects*, eds. Beckage, N. E., Thompson, S. M. & Federici, B. A. (Academic, San Diego), Vol. 1, pp. 189–225.
7. Webb, B. A. & Summers, M. D. (1992) *Experientia* **48,** 1018–1022.
8. Beckage, N. E. (1993) in *Parasites and Pathogens of Insects*, eds. Beckage, N. E., Thompson, S. M. & Federici, B. A. (Academic, San Diego), Vol. 1, pp. 25–57.
9. Beckage, N. E. (1993) *Receptor* **3,** 233–245.
10. Thompson, S. N. (1993) in *Parasites and Pathogens of Insects*, eds. Beckage, N. E., Thompson, S. M. & Federici, B. A. (Academic, San Diego), Vol. 1, pp. 125–144.

11. Tanaka, T. (1987) *J. Insect Physiol.* **33,** 413–420.
12. Beckage, N. E., Thompson, S. M. & Federici, B. A., eds. (1993) *Parasites and Pathogens of Insects* (Academic, San Diego).
13. Ratcliffe, N. A. (1993) in *Parasites and Pathogens of Insects*, eds. Beckage, N. E., Thompson, S. M. & Federici, B. A. (Academic, San Diego), Vol. 1, pp. 267–304.
14. Gupta, A. P. (1991) *Comparative Arthropod Morphology, Physiology, and Development* (CRC, Ann Arbor, MI).
15. Lackie, A. M. (1986) in *Immune Mechanisms in Invertebrate Vectors*, ed. Lackie, A. M. (Clarendon, Oxford), pp. 161–178.
16. Brehélin, M. & Zachary, D. (1986) in *Immunity in Invertebrates*, ed. Brehélin, M. (Springer, New York), pp. 36–48.
17. Brehélin, M. (1990) *Res. Immunol.* **141,** 935–938.
18. Edson, K. M., Vinson, S. B., Stoltz, D. B. & Summers, M. D. (1981) *Science* **211,** 582–583.
19. Theilmann, D. A. & Summers, M. D. (1986) *J. Gen. Virol.* **67,** 1961–1969.
20. Davies, D. H., Strand, M. R. & Vinson, S. B. (1987) *J. Insect Physiol.* **33,** 143–153.
21. Beckage, N. E. (1985) *Annu. Rev. Entomol.* **30,** 371–413.
22. Lawrence, P. O. & Lauzrein, B. (1993) in *Parasites and Pathogens of Insects*, eds. Beckage, N. E., Thompson, S. M. & Federici, B. A. (Academic, San Diego), Vol. 1, pp. 59–86.
23. Vinson, S. B. & Iwantsch, G. F. (1980) *Q. Rev. Biol.* **55,** 143–165.
24. Dover, B. A., Davies, D. H., Strand, M. R., Gray, R. S., Keeley, L. L. & Vinson, S. B. (1987) *J. Insect Physiol.* **33,** 333–338.
25. Theilmann, D. A. & Summers, M. D. (1988) *Virology* **167,** 329–341.
26. Vinson, S. B., Edson, K. M. & Stoltz, D. B. (1979) *J. Invertebr. Pathol.* **34,** 133–137.
27. Kitano, H. (1986) *J. Insect Physiol.* **32,** 369–375.
28. Taylor, T. & Jones, D. (1990) *Biochim. Biophys. Acta* **1035,** 37–43.
29. Jones, D. & Coudron, T. (1993) in *Parasites and Pathogens of Insects*, eds. Beckage, N. E., Thompson, S. M., & Federici, B. A. (Academic, San Diego), Vol. 1, pp. 227–244.
30. Rizki, R. M. & Rizki, T. M. (1991) *J. Exp. Zool.* **257,** 236–244.
31. Guillot, F. S. & Vinson, S. B. (1972) *J. Insect Physiol.* **18,** 1315–1321.
32. Ables, J. R. & Vinson, S. B. (1982) *Entomophaga* **26,** 453–458.
33. Stoltz, D. B., Guzo, B., Belland, E. R., Lucaroti, C. J. & MacKinnon, E. A. (1988) *J. Gen. Virol.* **69,** 903–907.
34. Webb, B. A. & Summers, M. D. (1990) *Proc. Natl. Acad. Sci. USA* **87,** 4961–4965.
35. Krell, P. J., Summers, M. D. & Vinson, S. B. (1982) *J. Virol.* **43,** 859–870.
36. Francki, R. I. B., Fauquet, C. M., Knudson, D. L. & Brown, F., eds. (1991) *Classification and Nomenclature of Viruses* (Springer, New York).
37. Krell, P. J. (1991) in *Viruses of Invertebrates*, ed. Kurstak, E. (Dekker, New York), pp. 141–177.
38. Blissard, G. W., Smith, O. P. & Summers, M. D. (1987) *Virology* **160,** 120–134.
39. Strand, M. R., McKenzie, D. I., Grassl, V., Dover, B. A. & Aiken, J. M. (1992) *J. Gen. Virol.* **73,** 1627–1635.
40. Fleming, J. G. W., Blissard, G. W., Summers, M. D. & Vinson, S. B. (1983) *J. Virol.* **48,** 74–78.
41. Blissard, G. W., Vinson, S. B. & Summers, M. D. (1986) *J. Virol.* **169,** 78–89.
42. Blissard, G. W., Fleming, J. G. W., Vinson, S. B. & Summers, M. D. (1986) *J. Insect Physiol.* **32,** 351–359.

43. Blissard, G. W., Theilmann, D. A. & Summers, M. D. (1989) *Virology* **169,** 78–89.
44. Theilmann, D. A. & Summers, M. D. (1987) *J. Virol.* **61,** 2589–2598.
45. Dib-Hajj, S. D., Webb, B. A. & Summers, M. D. (1993) *Proc. Natl. Acad. Sci. USA* **90,** 3765–3769.
46. Rodriguez, J. M., Yañez, R. J., Pan, R., Rodriguez, J. F., Salas, M. L. & Viñuela, E. (1994) *J. Virol.* **68,** 2746–2751.
47. Efstathiou, S., Lawrence, G. L., Brown, C. M. & Barrell, B. G. (1992) *J. Gen. Virol.* **73,** 1661–1671.
48. Thomson, B. J. & Honess, R. W. (1992) *J. Gen. Virol.* **73,** 1649–1660.
49. Upton, C., Macen, J. L., Wishart, D. S. & McFadden, G. (1990) *Virology* **179,** 618–631.
50. Liebhaber, S. A., Goossens, M. & Kan, Y. W. (1981) *Nature (London)* **290,** 26–29.
51. Ganz, T., Rayner, J. R., Valore, E. V., Rumolo, L., Ralmadge, K. & Fuller, F. H. (1989) *J. Immunol.* **143,** 1358–1365.
52. Padget, R. A., Grabowski, P. J., Konarska, M. M., Seiler, S. & Sharp, P. A. (1986) *Annu. Rev. Biochem.* **55,** 1119–1150.
53. Queen, C. & Baltimore, D. (1983) *Cell* **33,** 741–748.
54. Picard, D. & Schaffner, W. (1984) *Nature (London)* **307,** 80–82.
55. Lichtenstein, M., Keini, G., Cedar, H. & Bergnam, Y. (1994) *Cell* **76,** 913–923.
56. Whitelaw, C. B., Archibald, A. L., Harris, S., McClenaghan, M., Simons, J. P. & Clark, A. J. (1991) *Transgenic Res.* **1,** 3–13.
57. Bourbon, H.-M. & Amalric, F. (1990) *Gene* **88,** 187–196.
58. Frendewey, D., Barta, I., Gillespie, M. & Potashkin, J. (1990) *Nucleic Acids Res.* **18,** 2025–2032.
59. Sollner-Webb, B. (1993) *Cell* **75,** 403–405.
60. Liu, J. & Maxwell, E. S. (1990) *Nucleic Acids Res.* **18,** 6565–6571.
61. Henikoff, S., Keene, M. A., Fechtel, K. & Fristrom, J. W. (1986) *Cell* **44,** 32–42.
62. Andreadis, A., Gallego, M. E. & Nadal-Ginard, B. (1987) *Annu. Rev. Cell Biol.* **3,** 207–242.
63. Smith, C. W. J., Patton, J. G. & Nadal-Ginard, B. (1989) *Annu. Rev. Genet.* **23,** 527–577.
64. Olivera, B. M., Rivier, J., Clark, C., Ramilo, C. A., Corpuz, G. P., Abogadie, F. C., Mena, E. E., Woodward, S. R., Hillyard, D. R. & Cruz, L. J. (1990) *Science* **249,** 257–263.
65. Woodward, S. R., Cruz, L. J., Olivera, B. M. & Hillyard, D. R. (1990) *EMBO J.* **9,** 1015–1020.
66. Kimura, M. (1983) *The Neutral Theory of Molecular Evolution* (Cambridge Univ. Press, Cambridge, U.K.).
67. Hughes, A. L. & Nei, M. (1988) *Nature (London)* **335,** 167–170.
68. Hughes, A. L. & Nei, M. (1989) *Proc. Natl. Acad. Sci. USA* **86,** 958–962.
69. Nei, M. (1987) *Molecular Evolutionary Genetics* (Columbia Univ. Press, New York).
70. Von Heijne, G. (1986) *Nucleic Acids Res.* **14,** 4683–4690.
71. Linder, P., Lasko, P. F., Ashburner, M., Leroy, P., Nielsen, P. J., Nishi, K., Schnier, J. & Slonimski, P. P. (1989) *Nature (London)* **337,** 121–122.
72. Wasserman, D. A. & Steitz, J. A. (1991) *Nature (London)* **349,** 463–464.
73. Company, M., Arenas, J. & Abelson, J. (1991) *Nature (London)* **349,** 487–493.
74. Walker, J. E., Saraste, M., Runswick, M. J. & Gay, N. J. (1982) *EMBO J.* **1,** 945–951.
75. Pause, A. & Sonenberg, N. (1992) *EMBO J.* **11,** 2643–2654.

76. Schmid, S. R. & Linder, P. (1991) *Mol. Cell. Biol.* **11,** 3463–3471.
77. Stark, G. R., Debatisse, M., Guilutto, E. & Whal, G. M. (1989) *Cell* **57,** 901–908.
78. Murphy, P. M. (1993) *Cell* **72,** 823–826.
79. Gray, G. D. & Gill, H. S. (1993) *Int. J. Parasitol.* **23,** 485–494.
80. Thompson, N. S. (1986) *Comp. Biochem. Physiol.* **81,** 21–42.
81. Beckage, N. E. (1990) *UCLA Symp. Mol. Cell. Biol.* **112,** 497–515.
82. Palmer, J. D. & Shields, C. R. (1984) *Nature (London)* **307,** 437–440.
83. Kinoshita, Y., Ohnishi, N., Yama, Y., Kunisada, T. & Yamagishi, H. (1985) *Plant Cell Physiol.* **26,** 1401–1409.
84. Rogers, W. O. & Wirth, D. F. (1987) *Proc. Natl. Acad. Sci. USA* **84,** 565–569.
85. Hajduk, S. L., Klein, V. A. & Englund, P. T. (1984) *Cell* **36,** 483–492.
86. Betlelsen, A. H., Humayun, M. Z., Karfopoulos, S. G. & Rush, M. G. (1982) *Biochemistry* **21,** 2076–2085.
87. Sunnerhagen, P., Sjöberg, R.-M., Karlsson, A.-L., Lundh, L. & Bjursell, G. (1986) *Nucleic Acids Res.* **14,** 7823–7838.
88. Sunnerhagen, P., Sjöberg, R.-M. & Bjursell, G. (1989) *Somat. Cell Mol. Genet.* **15,** 61–70.
89. Fujimoto, S., Tsuda, T., Toda, M. & Yamagishi, H. (1985) *Proc. Natl. Acad. Sci. USA* **82,** 2072–2076.
90. Ruiz, J. C., Choi, K., Von Hoff, D. D., Roninson, I. G. & Wahl, G. M. (1989) *Mol. Cell. Biol.* **9,** 109–115.
91. Rush, M. G. & Misra, R. (1985) *Plasmid* **14,** 177–191.
92. Degroote, F., Pont, G., Micard, D. & Picard, G. (1989) *Chromosoma* **98,** 201–206.
93. Toda, M., Fujimoto, S., Iwasato, T., Takeshita, S., Tezuka, K., Ohbayashi, T. & Yamagishi, H. (1988) *J. Mol. Biol.* **202,** 219–231.
94. Toda, M., Hirama, T., Takeshita, S. & Yamagishi, Y. (1989) *Immunol. Lett.* **21,** 311–316.
95. Matsuoka, M., Yoshida, K., Maeda, T., Usuda, S. & Sakano, H. (1990) *Cell* **62,** 135–142.
96. von Schwedler, U., Jack, H.-M. & Wable, M. (1990) *Nature (London)* **345,** 452–456.
97. Gooding, L. R. (1992) *Cell* **71,** 5–7.
98. Guarino, L. A. (1990) *Proc. Natl. Acad. Sci. USA* **87,** 409–413.
100. Bigot, Y., Hamelin, M.-H., Capy, P. & Periquet, G. (1994) *Proc. Natl. Acad. Sci. USA* **91,** 3408–3412.
101. Van der Ploeg, L. H. T. (1987) *Cell* **51,** 159–161.
102. Vickerman, K. (1989) *Parasitology* **99,** S37–S47.
103. Fleming, J. G. W. (1992) *Annu. Rev. Entomol.* **37,** 401–425.
104. Li, W.-H. & Graur, D. (1990) *Fundamentals of Molecular Evolution* (Sinauer, Sunderland, MA).
105. Hughes, A. L. (1991) *Genetics* **127,** 345–353.
106. Hughes, A. L. (1994) *Parasitol. Today* **10,** 149–151.
107. Fitch, W. M., Leiter, J. M. E., Li, X. & Palese, P. (1991) *Proc. Natl. Acad. Sci. USA* **88,** 4270–4274.
108. Tanaka, T. & Nei, M. (1989) *Mol. Biol. Evol.* **6,** 447–459.
109. Smith, G. L. (1993) *J. Gen. Virol.* **74,** 1725–1740.
110. Soldevila, A. I. & Jones, D. (1994) *Insect Biochem. Mol. Biol.* **24,** 29–38.
111. Ferkovich, S. M., Greany, P. D. & Dillard, C. R. (1983) *J. Insect Physiol.* **24,** 933–942.
112. Cook, D. I., Stoltz, D. B. & Vinson, S. B. (1984) *Insect Biochem.* **14,** 45–50.

The Chemistry of Gamete Attraction: Chemical Structures, Biosynthesis, and (A)biotic Degradation of Algal Pheromones

WILHELM BOLAND

In 1971, Müller, Jaenicke, and colleagues (1) isolated the first pheromone of a marine brown alga. The compound was collected from laboratory cultures of fertile female gametophytes of the cosmopolitan brown alga *Ectocarpus siliculosus*. Soon after release, the originally motile female microgametes begin to settle on a surface and start to secrete a chemical signal. The biological function of this pheromone is the improvement of mating efficiency by attraction of the flagellated, motile males. The chemical structure of the signal compound was established as 6-(1*Z*)-(butenyl)cyclohepta-1,4-diene (ectocarpene; Figure 1 and Table 1). In the following years, this compound proved to be the progenitor of a whole series of C_{11} hydrocarbons, which are involved as signals in the sexual reproduction of brown algae. Up to now, 11 such pheromones (Table 1) but more than 50 stereoisomers of the parent compounds (cf. Figure 1) have been identified within the pheromone bouquets of >100 different species of brown algae (2–6). In the highly evolved orders Laminariales, Desmarestiales, and Sporochnales, the sexual pheromones first induce spermatozoid release from antheridia prior to attraction and fertilization (7, 8). Other species, in particular those from the genus *Dictyopteris*, produce large amounts of the same C_{11} hydrocarbons in their thalli. In the case of the Mediterranean phaeophyte *Dictyopteris membranaceae* (9), the compounds are released into the environment (10), and

Wilhelm Boland is chair of the Department of Bioorganic Chemistry at the Institut für Organische Chemie in Bonn, Germany.

TABLE 1 C_{11} and C_8 pheromones from marine brown algae

Chemical structures	Release/attraction (R/A)	Algal species	Chemical structures	Release/attraction (R/A)	Algal species
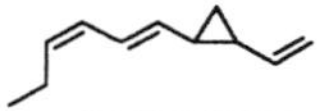 hormosirene	A	*Hormosira banksiii*	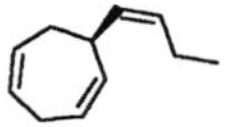 ectocarpene	A	*Ectocarpus spp.*
	A	*Durvillea spp.*		A	*Adenocystis utricularis*
	A	*Xiphophora spp.*		A	*Sphacelaria rigidula*
	A	*Scytosiphon lomentaria*			
	A	*Colpomenia perergrina*			
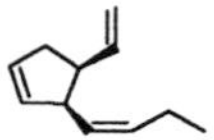 multifidene	A	*Cutleria multifida*	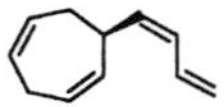 desmarestene	R/A	*Desmarestia spp.*
	R/A	*Chorda tomentosa*		A	*Cladostephus spongiosus*
	A	*Zonaria angustata*			
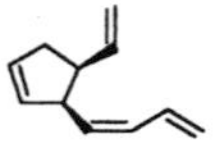 viridiene	A	*Syringoderma phinneyi*	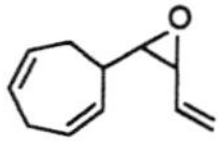 lamoxirene	R/A	*Laminaria spp.*
	R/A	*Desmarestia viridis*		R/A	*Alaria spp.*
				R/A	*Undaria pinnatifida*
				R/A	*Macrocystis pyrifera*
				R/A	*Nereocystis luetkeana*
				R/A	*many others*
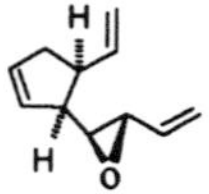 caudoxirene	R	*Perithalia caudata*	 cystophorene	A	*Cystophora siliquosa*
		Sporochnus radiciformis	 finavarrene	A	*Ascophyllum nodosum*
				A	*Sphaerotrichia divaricata*
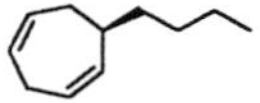 dictyotene	A	*Dictyota dichotoma*	 fucoserratene	A	*Fucus spp.*
	A	*Dictyota diemensis*			

References are mentioned in the text and in refs. 2–6.

here they may interfere with the communication systems of other brown algae (11) or act as feeding deterrents against herbivores (12). Thus, a single C_{11} hydrocarbon (Table 1) may have at least three well-defined biological functions: (*i*) synchronization of the mating of male and female cells by the controlled release of male spermatozoids, (*ii*) enhancement of the mating efficiency by attraction, and (*iii*) chemical defense of the plant due to the presence of high amounts of pheromones within and release from the thalli into the environment.

Interestingly, the occurrence of the C_{11} hydrocarbons is not limited to the order of the marine brown algae. The same compounds have been found in cultures of diatoms (13) or among the volatiles released during blooms of microalgae in freshwater lakes (14). Furthermore, there is a steadily increasing number of reports on the occurrence of C_{11} hydrocarbons in roots, leaves, blossoms, or fruits of higher plants (15–17), although no attempts have been made to attribute a specific biological function to the C_{11} hydrocarbons in these higher plants.

This review tries to summarize our present knowledge of the chemistry of the pheromones of marine brown algae. The biochemical aspects of signal perception and transduction as well as the species-specific recognition of male and female gametes will be not discussed. The major topics will be the demonstration of the structural diversity in this group of C_{11} hydrocarbons, the different modes of their biosynthesis in lower and higher plants, and, finally, their (a)biotic degradation in an aqueous environment.

BROWN ALGAE AND THEIR PHEROMONES

Since the isolation and characterization of ectocarpene as the first volatile pheromone of a brown alga, 10 additional bioactive compounds have been isolated from female gametes or eggs of brown algae (Table 1). The isolation technique exploits the volatility of the compounds by collecting them in a closed system with a continuously circulating stream of air, which is passed over a very small carbon trap (1.5–5 mg) (18). Solvent desorbtion of the carbon with CH_2Cl_2 or CS_2 ($\approx 30\ \mu l$) provides solutions of the volatiles suitable for analysis by combined gas chromatography/mass spectrometry. A complex pattern of volatiles, obtained from fertile female gynogametes of the Mediterranean phaeophyte *Cutleria multifida*, is shown in Figure 1 (19). The identification of the individual compounds follows from mass spectrometry and from comparison with synthetic reference compounds (2). The biological activity of the identified compounds is assayed by exposing microdroplets of a water immiscible, high density solvent with known concentrations of the compounds to male gametes in sea water (20). After 4 min in the dark, the

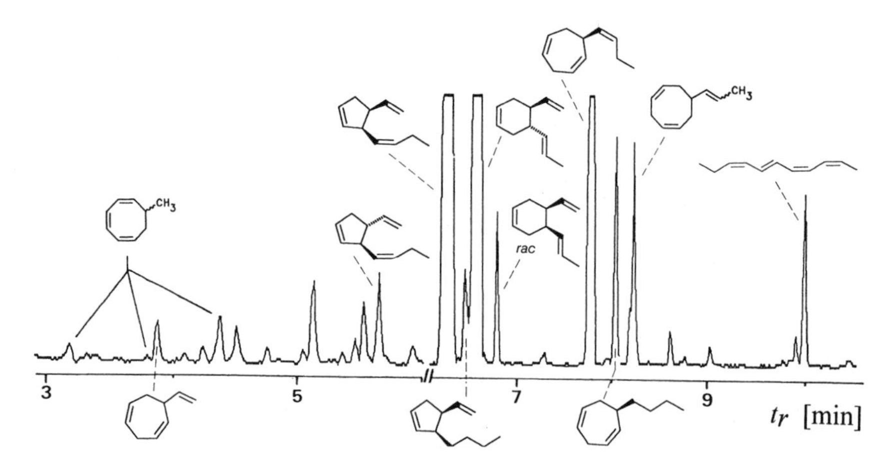

FIGURE 1 Gas chromatographic separation of the collected volatiles from fertile gynogametophytes of *C. multifida*. Conditions: fused silica column SE 30 (10 m × 0.32 mm); 40°C isotherm for 2 min, then at 10°C/min to 250°C; injection port: 250°C; detector: Finnigan ion trap, ITD 800; transfer line at 270°; electron impact (70 *e*V); scan range, 35–250 Da/sec. Due to the very low concentration of the low boiling point compounds, the total ion current in the range 3–6 min is enhanced by a factor of 6. Structures of the relevant hydrocarbons are given. Three configurational isomers appear for 7-methylcycloocta-1,3,5-triene owing to the high temperatures of the gas chromatographic separation.

population of the male gametes above the droplets of pure solvent and those containing the bioactive substances is documented by flash photography. For highly active pheromones, massive accumulation of males is observed in the range between 1 and 1000 pmol; the calculated values are valid for the saturated solvent/water interphase (21). Of the large number of the structurally very diverse compounds of Figure 1, the male gametes of *C. multifida* respond only to multifidene at a threshold level of 6.5 pmol and to ectocarpene at 900 pmol. The other compounds are virtually inactive. This also holds for the configurational isomers of multifidene, such as (*E*)-butenyl and 3,4-*trans*-disubstituted analogues, which on average exhibit only 1% the activity of the pheromone (21). Interestingly, even the activity difference between the natural (+)-(3*S*,4*S*)- and the synthetic (−)-(3*R*,4*R*)-multifidene is of the same order (22). Following this general sequence of isolation, characterization, synthesis, and bioassay, the pheromones of >100 species of brown algae have been determined (cf. Table 1) (3–6).

The ability of a compound to induce mass release of male gametes is approached by exposing mature male gametophytes of a species to particles of porous silica previously loaded with the test substances.

Quantitative data are available by placing droplets of an inert solvent with known concentrations of the test substance into close vicinity of the fertile gynogametophyte. For example, the mass release of male gametes of *Laminaria digitata* occurs within 8–12 sec at a threshold of ≈50 pmol of lamoxirene (8, 23, 24), and male gametes of *Perithalia caudata* are released within ≈10 sec, triggered by caudoxirene at concentrations down to 30 pmol (25).

Considering the large number of plant species and the limited number of only 11 different pheromones, it becomes obvious that these signals cannot be specific at the level of the species or even the genus. Moreover, ectocarpene, hormosirene, and dictyotene are typically present in most of the pheromone blends, and, hence, their presence may reflect nothing but a phylogenetic reminiscence of a biosynthetic pathway that converts a whole array of appropriate precursor molecules (see below) into olefinic hydrocarbons. The quantitative determination of the released volatiles from signaling females shows that females of *E. siliculosus* (26) release within 1 h ≈0.6 fmol of ectocarpene per individual; ≈75 fmol per h per egg of hormosirene was reported (27) as the initial rate of the secretory capacity of individual females of *Hormosira banksii.*

Much larger amounts of volatiles are present in thalli of several members of the genus *Dictyopteris*, but their production is obviously not linked to sexual events (ref. 28 and references cited therein). Interestingly, *Dictyopteris divaricata* from Japan (29) and *Dictyopteris zonarioides* from California (30, 31) produce sesquiterpenoids instead of C_{11} hydrocarbons. In the case of the Mediterranean *D. membranaceae*, the compounds are continuously released to the environment. The thalli of this alga (1 kg wet weight) release 10–50 mg of C_{11} hydrocarbons into seawater within 24 h (10). This exceeds by far the required threshold concentrations for chemotaxis, and one may suspect that the compounds can act as "mating disruptants" in the control of the habitat. This idea, however, awaits experimental confirmation. According to recent analyses, the occurrence of large amounts of C_{11} hydrocarbons in thalli of brown algae is not limited to the genus *Dictyopteris*. Two Fucales from the Red Sea—namely, *Sargassum asporofolium* and *Sargassum latifolium* (S. Fatallah and W.B., unpublished data)—contain, besides sesquiterpenes, ectocarpene and dictyotene as the major volatiles.

Recent advances in gas chromatographic separations of enantiomers allow precise determination of the enantiomeric purity of the algal pheromones. The *cis*-disubstituted cyclopentenes, such as multifidene, viridiene, and caudoxirene, are of high optical purity [≥95% enantiomeric excess (*e.e.*)] whenever they have been found (32, 33). The situation is different with the cyclopropanes and the cycloheptadienes, as shown in Table 2 and Figure 1. Hormosirene from female gametes or thalli of

TABLE 2 Enantiomer composition of hormosirene from secretions of female gametes or thalli of brown algae

Genus and species	Origin	Major enantiomer	*ee*, %
Dictyopteris acrostichoides	Sorrento, Australia	(−)-(1*R*,2*R*)	74.2
D. membranaceae	Villefranche, French Mediterranean	(−)-(1*R*,2*R*)	71.2
*D. prolifera**	Hikoshima, Japan	(−)-(1*R*,2*R*)	90.0
*D. undulata**	Hikoshima, Japan	(−)-(1*R*,2*R*)	92.0
*Analipus japonicus**	Muroran, Japan	(+)-(1*S*,2*S*)	66.0
*A. japonicus**	Akkeshi, Japan	(+)-(1*S*,2*S*)	90.0
Durvillaea potatorum	Sorrento, Australia	(−)-(1*R*,2*R*)	51.7
Haplospora globosa	Halifax, Nova Scotia	(+)-(1*S*,2*S*)	83.3
Hormosira banksii	Flinders, Australia	(−)-(1*R*,2*R*)	82.8
Xiphophora gladiata	Hobart, Tasmania	(−)-(1*R*,2*R*)	72.3
X. chondrophylla	Flinders, Australia	(−)-(1*R*,2*R*)	82.0

References are given in the text. Examples marked with an asterisk are taken from ref. 31.

brown algae of different geographic origin proved to be secreted always as a well-defined mixture of enantiomers (34). The enantiomeric composition of hormosirene from gametes of *Analipus japonicus* varies depending even on the locality within Japan (31). In contrast, the *e.e.* of hormosirene, present as a by-product (4–6%) in the pheromone bouquet of *E. siliculosus*, collected at various localities from all over the world, seems to remain constant (90% ± 5% *e.e.*; W.B. and D. G. Müller, unpublished data). If less common configurations of the pheromones are concerned, as, for example, in C_{11} hydrocarbons with (*E*)-butenyl or (*E,E*)-hexadienyl substituents, the optical purity is often very low. 6-(1*E*)-(Butenyl)cyclohepta-1,4-diene from the Australian *Dictyopteris acrostichoides* exhibits an *e.e.* of only 26% (10); the *cis*-disubstituted cyclohexene from the blend of *C. multifida* (Figure 1) is virtually racemic (19). It is tempting to assume that for marine brown algae the production of characteristic enantiomeric mixtures represents a simple means for individualization of the signal blends, although up to now there is no experimental confirmation for this hypothesis.

BIOSYNTHESIS OF C_{11} HYDROCARBONS IN HIGHER AND LOWER PLANTS (PHAEOPHYCEAE)

Given the absence of methyl branches and according to the suggestive positions of the double bonds within the two acyclic C_{11} hydrocarbons undeca-(1,3*E*,5*Z*)-triene and undeca-(1,3*E*,5*Z*,8*Z*)-tetraene, their origin from fatty acids is highly probable. In the case of higher plants, the

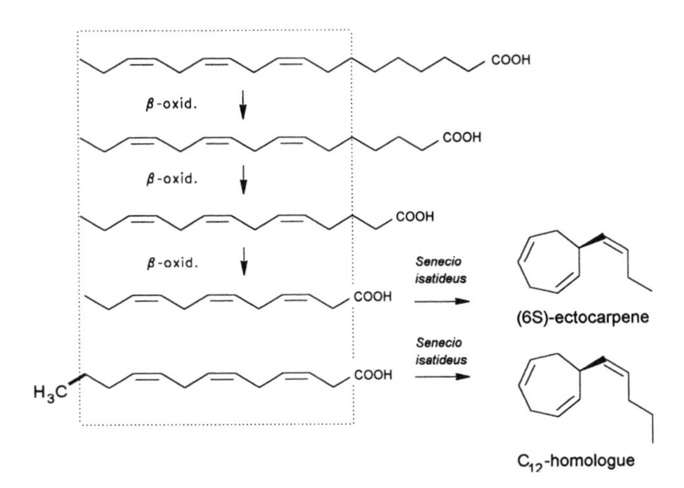

FIGURE 2 Biosynthesis of ectocarpene in the higher plant *S. isatideus* (Asteraceae). ω3 fatty acids and their degradation products (β-oxidation) comprise the structural elements for the biosynthesis of ectocarpene in the higher plant *S. isatideus.* The structurally related trideca-3,6,9-trienoic acid is metabolized by analogy into the C_{12} homoectocarpene.

precursors might be expected to come from the pool of the ω3 and ω6 fatty acids with a total of 18–12 carbon atoms (Figure 2). In marine brown algae the family of unsaturated C_{20} fatty acids provides another, potentially abundant source of suitable precursors (35).

Interestingly, the first biosynthetic experiments with [^{3}H]linolenic acid and the terrestrial plant *Senecio isatideus* (Asteraceae) as a model system for the biosynthesis of algal pheromones were unsuccessful. If, however, labeled dodeca-3,6,9-trienoic acid is administered, a rapid transformation into ectocarpene takes place (36). Nevertheless, the C_{12} acid ultimately derives from linolenic acid via three β-oxidations, since labeled tetradeca-5,8,11-trienoic acid, which requires only one β-oxidation, is converted into labeled ectocarpene albeit with very low efficiency (37).

Mechanistic insight into this process was obtained by administration of labeled trideca- or undeca-3,6,9-trienoic acid instead of the natural C_{12} precursor (Figure 2). In this case, the artificial ^2H_n metabolites can be analyzed by mass spectrometry without interference from the plants' own ^1H metabolites, since a homo- or norectocarpene is formed. The sequence of the oxidative decarboxylation/cyclization reaction proceeds without loss of ^2H atoms from the double bonds but with loss of a single ^2H atom from certain methylene groups of the precursor acids (Figure 3). If C(1) and a ^2H atom from C(5) of the labeled precursor is lost, finavarrene is the product of the reaction channel. If the methylene group

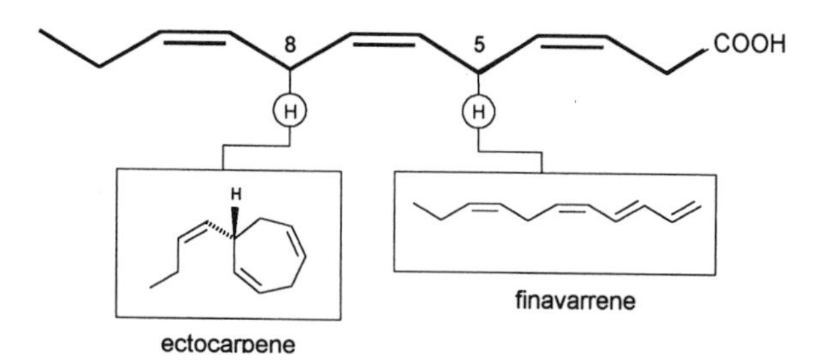

FIGURE 3 C_{11} hydrocarbons from dodeca-3,6,9-trienoic acid. Loss of C(1) and a single hydrogen from C(5) yield the acyclic hydrocarbon finavarrene. Decarboxylation and loss of a single hydrogen from C(8) results in (6*S*)-ectocarpene. No other hydrogen atoms are lost during the biosynthetic sequence.

at C(8) is involved, ectocarpene is formed with simultaneous loss of C(1), the latter probably as CO_2 (36). Administration of (8*R*)- or (8*S*)-[^{2}H]trideca-3,6,9-trienoic acids (38) yields (6*S*)-homoectocarpene with exclusive loss of the C(8) H_R. No intramolecular isotope effect is observed. The removal of the hydrogen atom from C(5) en route to finavarrene proceeds with the same side specificity. Considering the absolute configuration of ectocarpene as (6*S*) and taking into account the established preference of the enzyme(s) for the removal of the C(8) H_R, the course of the reaction from dodecatrienoic acid to the C_{11} hydrocarbon ectocarpene can be rationalized as depicted in Figure 4. The acid is assumed to fit into the active site of the enzyme in a U-shaped manner, exposing the C(8) H_R to a reactive functional group X. The sequence could be initiated by removal of a single electron from the carboxyl group. Subsequent decarboxylation would generate an allyl radical, which can interact with the C(6)═C(7) double bond generating a cyclopropyl intermediate. Final

FIGURE 4 Ectocarpene as the product of a [3.3]-sigmatropic rearrangement. The fatty acid accommodates to the active center of the enzyme in a U-shaped fashion. Decarboxylation in conjunction with loss of the C(8) H_R hydrogen atom yields, after cyclization between C(4) and C(6) of the precursor, the thermolabile (1*S*,2*R*)-cyclopropane. A subsequent spontaneous [3.3]-sigmatropic rearrangement (Cope rearrangement) proceeds via the *cis–endo* transition state and yields (6*S*)-ectocarpene.

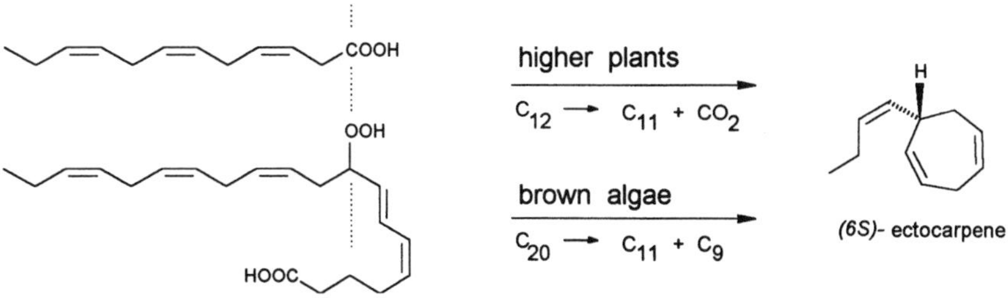

FIGURE 5 Biosynthesis of C_{11} hydrocarbons in higher and lower plants. The similar pattern of functionalization of dodeca-3,6,9-trienoic acid and 9-HPEPE is shown.

removal of the C(8) H_R as a radical by the active center of the enzyme terminates the sequence, and the disubstituted cyclopropane (1*S*,2*R*) is released as the first product. The compound is thermolabile and rearranges via a *cis–endo* transition state to (6*S*)-ectocarpene.

In contrast to the terrestrial plant *S. isatideus*, female gametes of the marine brown algae (model system: female gametes of *E. siliculosus*) do not utilize dodeca-3,6,9-trienoic acid for production of the C_{11} hydrocarbons. Instead, the marine plants exploit the pool of unsaturated C_{20} acids (20:4 → 20:6) for the production of their pheromones. [$^2H_8$]Arachidonic acid is very effectively transformed into 6-[$^2H_4$]butylcyclohepta-1,4-diene (dictyotene) by a suspension of female gametes of *E. siliculosus*. A synthetic sample of [$^2H_6$]nonadeca-8-11,14,17-tetraenoic acid, which can be thought of as a 20-noranalogue of icosa-5,8,11,14,17-pentaenoic acid, gives the corresponding norectocarpene, together with a labeled norfinavarrene, in high yield (39, 40). Since the icosanoids are not cleaved to unsaturated C_{12} precursors, their primary functionalization is assumed to be achieved by a 9-lipoxygenase yielding 9-hydroperoxyicosa-(5*Z*,7*E*,11*Z*,14*Z*,17*Z*)-pentaenoic acid (9-HPEPE), which mimics the functionalization pattern of dodeca-3,6,9-trienoic acid (Figure 5). Assuming a homolytic cleavage of the hydroperoxide (41) as shown in Figure 6, once again, the disubstituted cyclopropane will be released from the active center as a thermolabile intermediate. Although this mechanistic hypothesis has not yet been experimentally confirmed, the concept nevertheless provides a valuable platform for the systematic derivation of all the known C_{11} hydrocarbons from the fatty acid precursors. It is also conceivable that the oxidative decarboxylation/cyclization of dodeca-3,6,9-trienoic acid in higher plants could proceed via a peroxy acid.

The unusual stereochemistry of the acyclic undeca-(2*Z*,4*Z*,6*E*,8*Z*)-tetraene (giffordene; cf. Figure 1), the major volatile from the brown alga *Giffordia mitchellae*, follows from the same concept (40, 42). If the 9-HPEPE

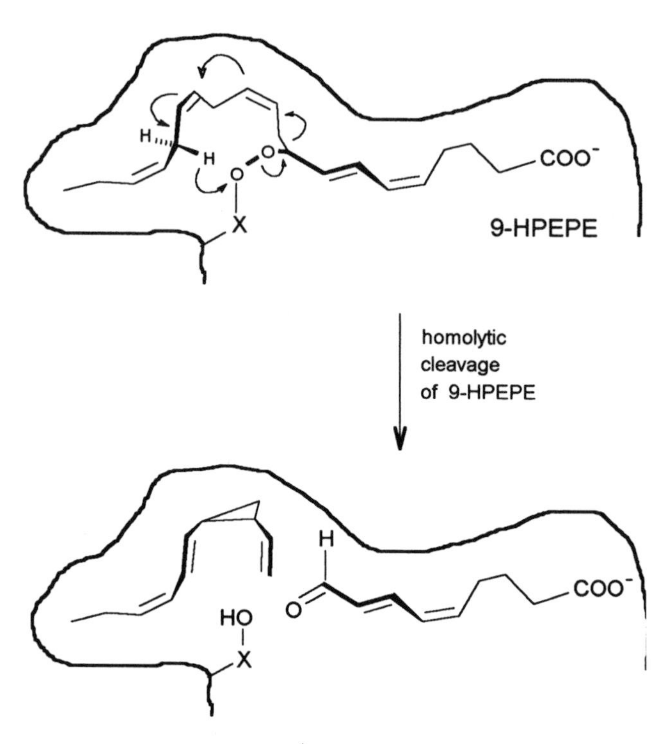

FIGURE 6 Speculative mechanism of C_{11} hydrocarbon biosynthesis from fatty acid hydroperoxides in algae. Homolytic cleavage of the hydroperoxide is assumed to give an allyl radical, which cyclizes to the thermolabile (1*S*,2*R*)-cyclopropane. The sequence is terminated by transfer of a hydrogen radical from C(16) to the -X-O˙ function. The cyclopropane rearranges to (6*S*)-ectocarpene as shown in Figure 4.

precursor is forced by the enzyme into an appropriate cisoid conformation, homolytic cleavage of the peroxide generates the thermolabile undeca-(1,3*Z*,5*Z*,8*Z*)-tetraene. The latter suffers a spontaneous [1.7]-hydrogen shift, and it is the helical transition state structure of this antarafacial hydrogen shift that accounts for the observed stereochemistry of the product (Figure 7*a*). The concept is strongly supported by the simultaneous occurrence of small traces of the more stable (1,3*Z*,5*Z*)-undecatriene in the same bouquet. A low temperature synthesis (−30°C) of the postulated undeca-(1,3*Z*,5*Z*,8*Z*)-tetraene has been developed, and from the kinetic data a half-life of ≈2.5 h is calculated for this compound under the conditions of the natural environment (18°C) (G. Pohnert and W.B., unpublished data). The activation energy of this [1.7]-hydrogen shift ($E_a = 67.4$ kJ·mol^{-1}; $\Delta S_{298} = -91.9$ J·mol^{-1}·K^{-1}) is considerably

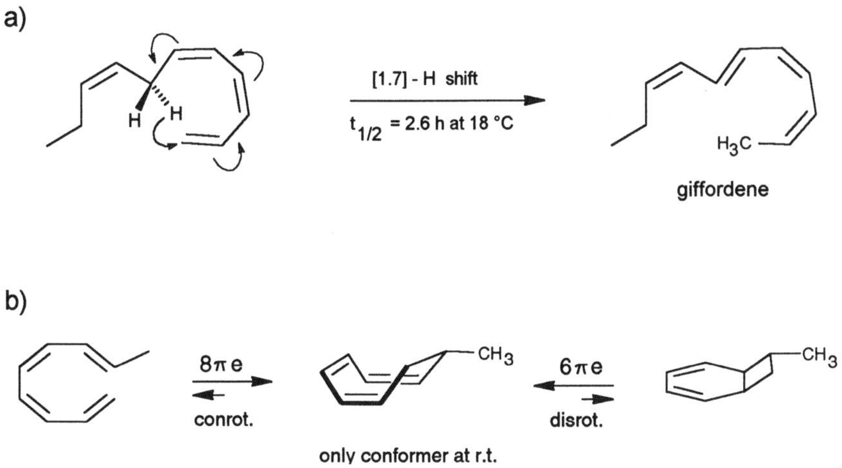

a)

H H

[1.7] - H shift

$t_{1/2}$ = 2.6 h at 18 °C

H_3C

giffordene

b)

CH_3

8π e

conrot.

only conformer at r.t.

6π e

disrot.

CH_3

FIGURE 7 Pericyclic reactions in the biosynthesis of giffordene and 7-methylcyclooctatriene. (*a*) The [1.7]-hydrogen shift of the thermolabile undeca-(1,3*Z*,5*Z*,8*Z*)-tetraene generates undeca-(2*Z*,4*Z*,6*E*,8*Z*)-tetraene (giffordene), the major product of the brown alga *G. mitchellae.* (*b*) The thermolabile nona-(1,3*Z*,5*Z*,8*E*)-tetraene cyclizes at ambient temperature rapidly to 7-methylcycloocta-1,3,5-triene. At ambient temperature, the bicyclic isomer does not contribute to the equilibrium (requires ≥80°C).

lower than that of the well studied [1.7]-hydrogen shift from previtamin D_3 to vitamin D_3 (43). Both of these pericyclic reactions do not appear to be catalyzed by enzymes.

Yet another pericyclic reaction may account for the biosynthesis of 7-methylcycloocta-1,3,5-triene, present at a trace level in the hydrocarbon blend from *C. multifida* (Figure 1). If a nona-(1,3*Z*,5*Z*,7*E*)-tetraene were produced by homolytic cleavage of a suitable fatty acid hydroperoxide, this acyclic olefin should readily undergo an $8\pi e$ electrocyclic ring closure (Figure 7*b*). The same reaction has been postulated previously as the key step within the biosynthetic sequence en route to the endriandric acids from the Australian tree *Endriandra introrsa* (45). Again, a low temperature synthesis (−30°C) of the nona-(1,3*Z*,5*Z*,7*E*)-tetraene allowed the determination of the kinetic data. The half-life of the acyclic precursor is limited to a few minutes at ambient temperature, and the activation energy is significantly lower (E_a = 59.4 kJ·mol^{-1}; ΔS_{273} = −89.7 J·mol^{-1}·K^{-1}) than those reported for the electrocyclization of a series of isomeric decatetraenes (44). The data follow a predicted trend based on theoretical calculations (46). In contrast to the three signals that were obtained by gas chromatographic analysis at higher temperature

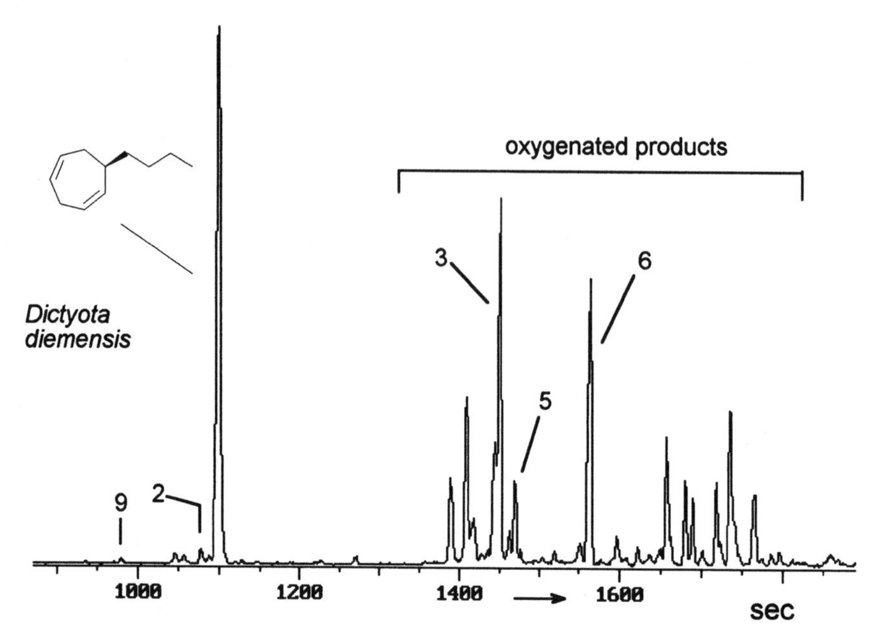

FIGURE 8 Gas chromatographic separation of the volatiles of *D. diemensis* egg extracts (47). Conditions: fused silica column OV 1 (10 m × 0.32 mm); 50°C isotherm for 2 min, then at 10°C/min to 250°C; injection port: 250°C; detector: Finnigan ion trap, ITD 800; transfer line at 270°C; electron impact (70 eV); scan range, 35–250 Da/sec. For identity of numbered compounds refer to Figure 9.

(Figure 1), ^{1}H NMR studies show that at ambient temperature, the monocyclic 7-methylcyclooctatriene is present as the only isomer (no evidence for an equilibrium with a bicyclic cyclohexadiene).

(A)BIOTIC DEGRADATION OF ALGAL PHEROMONES

The degree and extent of the abiotic degradation of the pheromones becomes immediately obvious when samples from algae releasing cyclohepta-1,4-dienes (such as dictyotene or ectocarpene) are collected. For example, volatiles collected from cultures of fertile gynogametophytes of *Dictyota diemensis* exhibit, besides dictyotene, a complex pattern of oxygenated compounds, as shown in Figure 8 (47). Since the concentrations of the oxygenated products are even lower than that of the genuine pheromone, the enrichments from natural sources can not be used for an exhaustive structure elucidation.

However, the pattern of Figure 8 can be perfectly mimicked by a chemical model system with dictyotene as the substrate and iodosylbenzene (Ph-I=O)/manganese tetraphenylporphyrin (TPPMn) as the oxidant (48). An aqueous system, containing huminic acids and traces of Cu(I) (49), gives similar results (50). Since these abiotic degradations can be carried out on a gram scale, the by-products shown in Figure 8 become available as pure compounds after extensive chromatographic separations. Up to now, more than 20 individual compounds have been characterized (50). Their structures fit into a general scheme of an oxidative sequence starting with a pentadienyl radical as depicted in Figure 9. Once generated, the radical reacts at all possible positions with oxygen and yields the three isomeric alcohols as the primary products. Further oxidation of the alcohols provides the dihydrotropones, which were first isolated from the two brown algae *Dictyopteris australis* and *Dictyopteris plagiogramma* (28, 51). Elimination of water generates the alkyltropilidenes. If singlet oxygen is involved, the corresponding hydroperoxides will be formed. Their subsequent decomposition may be responsible for the formation of the ketoepoxide and for the fragmentation of the carbon framework. The isomeric butylbenzaldehydes and the substituted furane fit into the same reaction channel. The butylbenzenes result from decarbonylation of the alkylbenzaldehydes. Butylbenzene and the alkyltropilidenes are remarkably attractive for male gametes of *E. siliculosus*; the alcohols and ketones are not. Owing to their ability to act as Michael acceptors, the dihydrotropones and tropones, which represent the major degradation products of the cycloheptadienes, may be important as chemical defenses (e.g., feeding deterrents) of those brown algae that are capable of synthesizing them. Similar degradative routes can be expected for all the other C_{11} hydrocarbons, and it seems likely that chemical model systems like Ph-I=O/TPPMn can provide all the required reference substances.

Exploratory experiments with dictyotene and suspensions of male gametes of *E. siliculosus* showed a significantly enhanced production of the 6-butylcyclohepta-2,4-dienol and its isomers. This indicates that a biological degradative pathway does exist and that this pathway follows the same oxidative sequence as the abiotic route. However, final conclusions about the biotic contribution to the pheromone transformation cannot be drawn before careful analysis of the degree of enantioselectivity of the biotic reaction.

SUMMARY

Female gametes of marine brown algae release and/or attract their conspecific males by chemical signals. The majority of these compounds are unsaturated, nonfunctionalized acyclic, and/or alicyclic C_{11} hydrocarbons. Threshold concentrations for release and attraction are generally

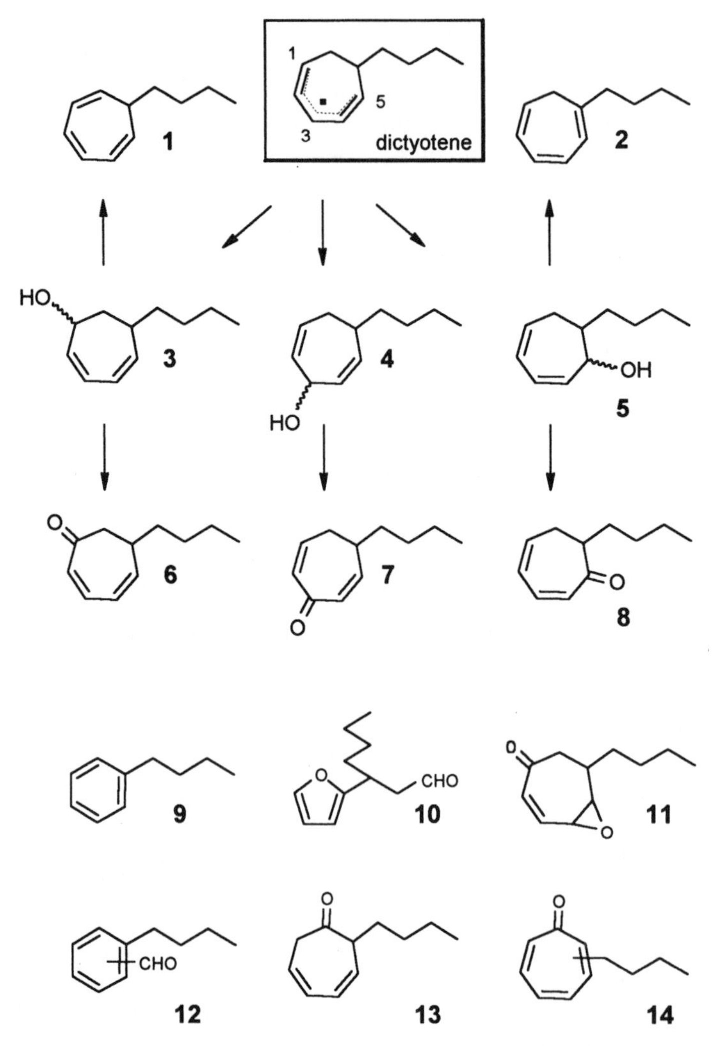

FIGURE 9 Oxidative degradation of dictyotene by TPPMn/Ph-I═O. Compounds **1–14** are isolated products of an oxidative degradation of dictyotene by a radical pathway induced by the system TPPMn/Ph-I═O. Most of these compounds are present among the oxygenated derivatives of dictyotene from natural sources (cf. Figure 8).

observed in the range of 1–1000 pmol. The blends may contain various configurational isomers of the genuine pheromones as well as mixtures of enantiomers. Higher plants produce the C_{11} hydrocarbons from dodeca-3,6,9-trienoic acid; brown algae exploit the family of icosanoids for biosynthesis of the same compounds. The biosynthetic routes comprise several spontaneously occurring pericyclic reactions such as [3.3]-sigmatropic rearrangements, [1.7]-hydrogen shifts, and electrocyclic ring closures. All pheromones are (a)biotically degraded by ubiquitous oxidative pathways involving singlet oxygen or hydroxyl radicals, which may be produced through the agency of heavy metals, huminic acids, or light.

REFERENCES

1. Müller, D. G., Jaenicke, L., Donike, M. & Akintobi, T. (1971) *Science* **171,** 815–817.
2. Jaenicke, L. & Boland, W. (1982) *Angew. Chem. Int. Ed. Engl.* **94,** 643–653.
3. Maier, I. & Müller, D. G. (1986) *Biol. Bull.* **170,** 145–175.
4. Boland, W. (1987) *Biol. Unserer Zeit* **17,** 176–185.
5. Jaenicke, L. (1988) *Bot. Acta* **101,** 149–159.
6. Maier, I. (1993) *Plant Cell Environ.* **16,** 891–907.
7. Maier, I. & Müller, D. G. (1982) *Protoplasma* **113,** 137–143.
8. Maier, I. (1987) in *Algal Development (Molecular and Cellular Aspects),* eds. Wiessner, W., Robinson, D. G. & Starr, R. C. (Springer, Berlin), pp. 66–74.
9. Boland, W. & Müller, D. G. (1987) *Tetrahedron Lett.* **28,** 307–310.
10. Wirth, D., Fischer-Lui, I., Boland, W., Icheln, D., Runge, T., König, W. A., Phillips, J. & Clayton, M. (1992) *Helv. Chim. Acta* **75,** 734–744.
11. Boland, W., Marner, F.-J., Jaenicke, L., Müller, D. G. & Fölster, E. (1983) *Eur. J. Biochem.* **134,** 97–103.
12. Hay, M. E., Duffy, J. E., Fenical, W. & Gustafson, K. (1988) *Mar. Ecol. Prog. Ser.* **48,** 185–192.
13. Derenbach, J. B. & Pesandeo, D. (1986) *Mar. Chem.* **19,** 337–432.
14. Jüttner, F. & Wurster, K. (1984) *Limnol. Oceanogr.* **29,** 1322–1324.
15. Bohlmann, F., Zdero, C., Berger, A., Suwita, A., Mahanta, P. & Jeffrey, C. (1979) *Phytochemistry* **18,** 79–93.
16. Boland, W., Jaenicke, L. & Brauner, A. (1982) *Z. Naturforsch.* **37C,** 5–9.
17. Kollmannsberger, H. & Berger, R. G. (1992) *Chem. Mikrobiol. Technol. Lebensm.* **14,** 81–86.
18. Grob, K. & Zürcher, F. (1976) *J. Chromatogr.* **117,** 285–294.
19. Keitel, J., Fischer-Lui, I., Boland, W. & Müller, D. G. (1990) *Helv. Chim. Acta* **73,** 2101–2112.
20. Müller, D. G. (1976) *Z. Pflanzenphysiol.* **80,** 120–130.
21. Boland, W., Jakoby, K., Jaenicke, L. & Müller, D. G. (1980) *Z. Naturforsch.* **36C,** 262–271.
22. Boland, W., Jaenicke, L. & Müller, D. G. (1981) *Justus Liebigs Ann. Chem.* 2266–2271.
23. Müller, D. G., Gassmann, G. & Lüning, K. (1979) *Nature (London)* **279,** 430–431.
24. Marner, F.-J., Müller, B. & Jaenicke, L. (1984) *Z. Naturforsch.* **39C,** 689–691.
25. Müller, D. G., Boland, W., Becker, U. & Wahl, T. (1988) *Biol. Chem. Hoppe-Seyler* **369,** 655–659.

26. Müller, D. G. & Schmid, C. E. (1988) *Biol. Chem. Hoppe-Seyler* **369,** 647–653.
27. Maier, I. & Clayton, M. N. (1993) *Bot. Acta* **106,** 344–349.
28. Moore, R. E. (1977) *Acc. Chem. Res.* **10,** 40–47.
29. Kajiwara, T., Hatanaka, A., Kodama, K., Ochi, S. & Fujimura, T. (1991) *Phytochemistry* **30,** 1805–1807.
30. Fenical, W. & Sims, J. J. (1972) *Phytochemistry* **11,** 1161–1163.
31. Kajiwara, T., Kodama, K. & Hatanaka, A. (1993) in *Volatile Attractants from Plants,* ACS Symposium Series 525 (Am. Chem. Soc., Washington, DC), pp. 103–120.
32. Boland, W., König, W. A. & Müller, D. G. (1989) *Helv. Chim. Acta* **72,** 1288–1292.
33. Wirth, D., Boland, W. & Müller, D. G. (1992) *Helv. Chim. Acta* **75,** 751–758.
34. Boland, W., Flegel, U., Jordt, G. & Müller, D. G. (1987) *Naturwissenschaften* **74,** 448–449.
35. Schmid, C. E., Eichenberger, W. & Müller, D. G. (1991) *Biol. Chem. Hoppe-Seyler* **372,** 540–544.
36. Boland, W. & Mertes, K. (1985) *Eur. J. Biochem.* **147,** 83–91.
37. Stratmann, K. (1992) Ph.D thesis (University of Karlsruhe, Karlsruhe, F.R.G.).
38. Neuman, C. & Boland, W. (1990) *Eur. J. Biochem.* **191,** 453–459.
39. Stratmann, K., Boland, W. & Müller, D. G. (1992) *Angew. Chem. Int. Ed. Engl.* **31,** 1246–1248.
40. Stratmann, K., Boland, W. & Müller, D. G. (1993) *Tetrahedron* **49,** 3755–3766.
41. Gardner, H. W. (1991) *Biochim. Biophys. Acta* **1084,** 221–239.
42. Boland, W., Jaenicke, L. & Müller, D. G. (1987) *Experientia* **43,** 466–467.
43. Hanewald, K. H., Rappoldt, M. P. & Roborgh, J. R. (1961) *Rec. Trav. Chim. Pays-Bas* **80,** 1003–1014.
44. Huisgen, R., Dahmen, A. & Huber, H. (1967) *J. Am. Chem. Soc.* **89,** 7130–7131.
45. Bandaranayake, W. M., Banfield, J. E. & Black, D. S. C. (1980) *J. Chem. Soc. Chem. Commun.* 902–903.
46. Thomas, B. E., Evanseck, J. D. & Houk, K. N. (1993) *J. Am. Chem. Soc.* **115,** 4165–4169.
47. Phillips, J. A., Clayton, M. N., Maier, I., Boland, W. & Müller, D. G. (1990) *Phycologia* **29,** 367–379.
48. McMurry, T. J. & Groves, J. T. (1986) in *Cytochrome P-450, Structure, Mechanism, and Biochemistry,* ed. Ortiz de Montellano, P.R. (Plenum, New York), pp. 1–28.
49. Mopper, K. & Zhou, X. (1990) *Science* **250,** 661–662.
50. Erbes, P. (1992) Ph.D. thesis (University of Karlsruhe, Karlsruhe, F.R.G.).
51. Moore, R. E. & Yost, G. (1973) *J. Chem. Soc. Chem. Commun.* 937–938.

The Chemistry of Sex Attraction

WENDELL L. ROELOFS

Mate attraction involves multiple sensory modalities with almost as many variations as there are numbers of species. It broadly includes pheromones of single-celled organisms to those involved in uniting mates over distances of several kilometers, elaborate plumages and songs in birds, croaks and chucks in frogs, electrical discharges in fish, etc. In insects, a pattern of light flashes from a firefly or of chirps from a cricket are conspicuous signals used to attract mates, but the most common form of long-distant sex attraction in insects (and many other animal species) involves invisible odors (chemicals!). One interesting exception to invisibility is when a visible aerosol of pheromone is produced to attract a mate (Figure 1).

The phenomenon of virgin female moths attracting large numbers of male moths was noted long before it was understood that a trail of chemicals could be so effective over great distances. In 1882 J. A. Lintner, the first state entomologist in New York, described (1) a spectacle of 50 large male *Promethea* moths being attracted to a female moth placed in his office window, which in turn attracted a large crowd of people on the sidewalk. He not only acknowledged the role of chemicals in this process and the existence of "smell organs" but also foresaw the potential use of these chemicals for insect control. He writes (1):

Wendell Roelofs is Liberty Hyde Bailey Professor of Insect Biochemistry at Cornell University, Geneva, New York.

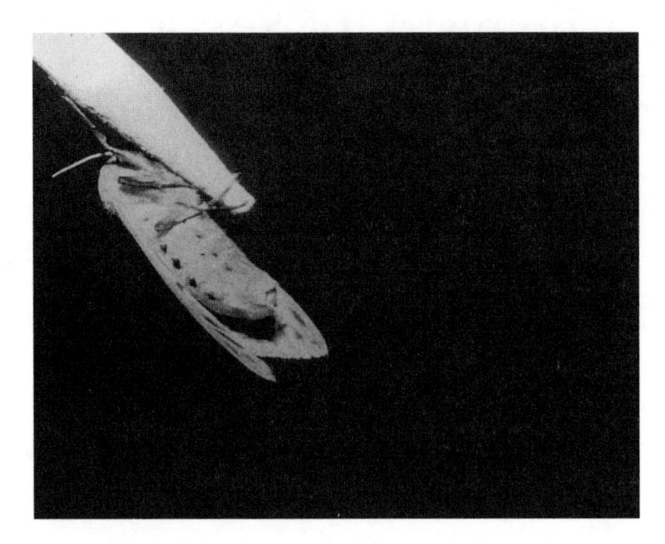

FIGURE 1 Female arctiid moth in a wind tunnel. Pheromone emission as a visible stream of liquid droplets is shown. Photo by S. Krasnoff.

It is the existence of these organs that many of the Families of Lepidoptera, especially among the moths, owe the continuance of the species. They operate in the association of the sexes, with such an irrestible and far-reaching force as to exceed our conception and to be marvelous to our eyes. . . . Can not chemistry come to the aid of the economic entomologist in furnishing at moderate cost, the odorous substances needed? Is the imitation of some of the more powerful animal secretions impracticable?

This statement sums up much of the research that has been conducted on the chemistry of sex attraction in insects over the past few decades. Research efforts have focused on the overt chemical signals used in the mating process, with the driving force and financial backing for much of the research due to the potential for use of synthetic pheromone chemicals in pest control programs. This effort has resulted in the identification of sex attractants for >1600 insect species from >90 families in nine orders, with an emphasis on Lepidoptera (2). The variety and complexity of chemical structures (>300 reported structures) observed among the various insect orders attest to the insect's amazing ability to sequester and synthesize unique structures and blends, mainly composed of acetogenins and mevalogenins. These essential signaling chemicals of the "sex attractant" communication system are interesting but represent only the more obvious chemicals involved in complex systems of synchronous emitters and receivers. Rather than review all data on this enormous topic, this paper will discuss some of the chemistry involved in various aspects of the sex pheromone communication system in insects.

OVERT CHEMICAL SIGNALS

The sex pheromone communication system basically involves the release of specific chemicals from a pheromone producer (emitter), the transmission of these chemicals in the environment to a receiver, and the processing of these signals to mediate appropriate behavioral responses in the receiver. The chemicals transmitted downwind have been the most obvious targets for characterization. The code was first broken with the publication in 1959 (3) of the sex pheromone for the domesticated silkworm *Bombyx mori* after extraction of a half million female silkworm pheromone glands and 30 years of classical chemical analyses. The pheromone was found to be (*E*10, *Z*12)-hexadecadien-1-ol, which was called bombykol. This work showed that there was nothing magical about the communication system, and chemists around the world were "attracted" to this area of research on insect pheromones.

Rapid Identification of Pheromone Components with the Electroantennogram

Research on the isolation and identification of various chemical signals used for mate attraction in insects has progressed through stages of improving technologies [e.g., capillary gas/liquid chromatography (GLC), airborne collectors, flight-tunnel bioassays, and powerful instrumental analyses on microsamples]. One of the most sensitive instruments used throughout the past two decades has been the electroantennogram (EAG) apparatus, which utilizes the male's antenna as a finely tuned detector for active materials. Schneider (4, 5) was the pioneer for setting up the EAG to carry out electrophysiological experiments on olfaction in insects. The sensitivity and specificity of the male's antenna to its own pheromone components made it a powerful tool in assaying for pheromone components and in predicting their structures (6). GLC retention times on nonpolar and polar columns of active compounds in crude female gland extracts could be determined quickly by collecting the GLC effluent in a separate capillary tube each minute and puffing air through each tube and across an isolated male antenna in the EAG setup. Depolarization of the isolated antenna in response to each puff is amplified and recorded or observed with an oscilloscope. The amplitude of response to active compounds was found to correlate to the frequency of generated nerve impulses, and thus, pheromone components in the extract elicited responses with the highest amplitudes.

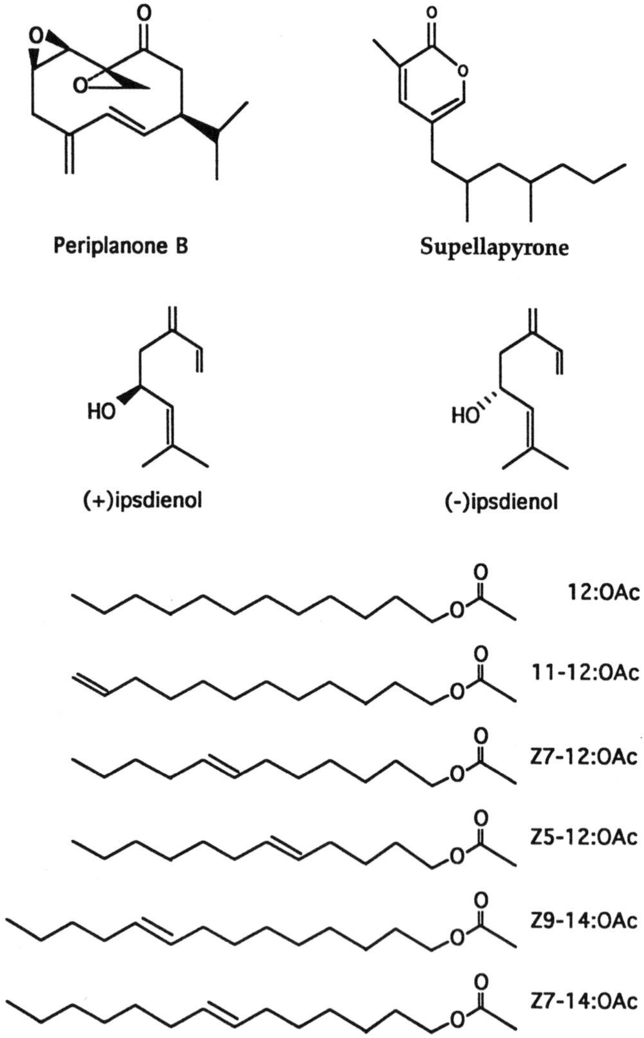

FIGURE 2 Pheromone structures of the American cockroach (periplanone B), the brownbanded cockroach (supellapyrone), bark beetles (ipsdienol enantiomers), and the cabbage looper moth (six acetates).

The EAG technique as a bioassay tool for active fractions has been used in the identification of sex pheromones of many insect species in several orders and remains as a key factor in the identification of pheromones. In the recent identification of the brownbanded cockroach pheromone supellapyrone (7) (Figure 2), the EAG technique was used throughout the entire process of isolating and purifying active material from 12,000 female roaches. Additionally, the EAG technique was used on many moth species to predict the positions and configuration of double bonds in long-chain fatty acetates, aldehydes, and alcohols. GLC retention time data was obtained to determine the carbon-chain length and functional group, and then every monounsaturated compound from the suggested library (e.g., monounsaturated 12-carbon acetates) was puffed across the antenna to determine relative response amplitudes. In many cases it was easy to determine that the main pheromone component was (*Z*8)- or (*Z*9)- or (*E*10)-dodecenyl acetate, etc. In the first use of the EAG for pheromone identification, the highly pursued structure of the codling moth sex pheromone was determined in a short time with only a few moths (8). The GLC retention time data indicated a conjugated 12-carbon alcohol, and responses from the library of monounsaturated 12-carbon alcohols showed that the antenna responded quite specifically to the (*E*8), the (*E*9), and the (*E*10) isomers in that series. These data indicated an (*E*8, *E*10) conjugated system. Synthesis and EAG analyses of all four 8,10 geometrical isomers showed that the *E,E* isomer produced the greatest antennal responses of all test compounds. Behavioral studies proved it to be a very potent sex attractant for this species, and it was later isolated from the female pheromone gland.

Complex Structures vs. Specific Blends

A mating communication signal wafting through a "noisy" chemical background in the environment must be specific for the receiver. One answer to this problem is to make a chemical that is so complex that it is not duplicated in nature and so imparts great species specificity. Single complex chemicals indeed have been found that elicit the full repertoire of behavioral responses at very low biologically meaningful concentrations. The cockroach sex pheromones (Figure 2) periplanone B for the American cockroach *Periplaneta americana* (9) and supellapyrone for the brownbanded cockroach *Supella longipalpa* (7) are examples of such chemicals that alone can attract conspecific males from a distance with the release of only hundreds of molecules.

Although a number of species utilize complex chemicals containing chiral centers and may, in fact, produce only one stereoisomer, the behavior of the responding male in some species is not affected by the presence of the other isomers and a racemic blend is as active as the pure isomer. However, other cases have been found in which a distinct chemical is found to be used by an insect species, but it is also part of the mating communication system of other related species. Species specificity then can be effected by utilizing a specific blend of the enantiomers (10). An example of this is with the two bark beetle species that utilize

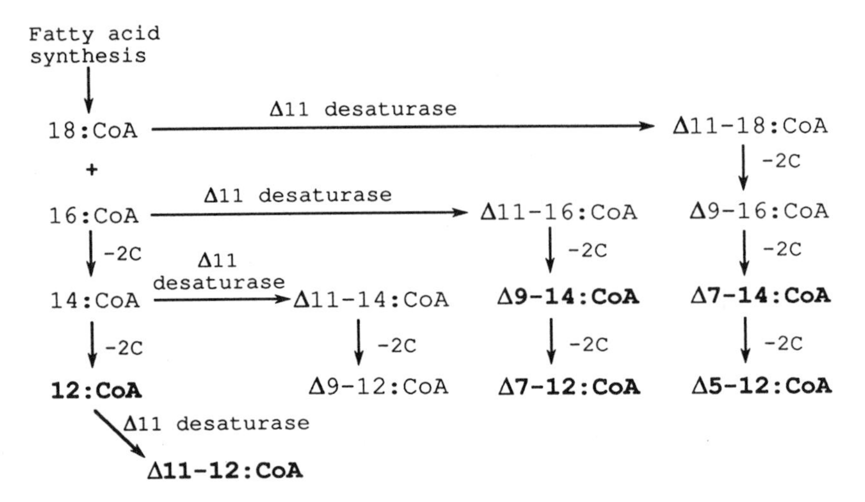

FIGURE 3 Pheromone biosynthetic pathways commonly used in moth sex pheromone glands to produce precursors for specific blends of acetates, alcohols, or aldehydes. Cascades of precursors are produced by combinations of unique Δ^{11}-desaturases and limited chain-shortening steps. The six precursors for the cabbage looper blend (Figure 2) are in boldface type.

ipsdienol. The western species *Ips pini* produces only the (−)-enantiomer, whereas the eastern species *Ips pini pini* utilizes a 65% (+)/35% (−) mixture (11) (Figure 2).

In contrast to pheromones that involve single complex compounds, many moth species have been found to utilize a specific blend of relatively simple fatty acid-derived compounds. It appears that the evolution of a unique enzyme, Δ^{11}-desaturase, used in combination with 2-carbon chain-shortening reactions (Figure 3) has allowed moth species to produce a variety of unsaturated acetates, aldehydes, and alcohols that can be combined in almost unlimited blends to impart species specificity. For example, biosynthetic precursors for the six-component pheromone blend of acetates for the cabbage looper moth (12) (Figure 2) can be determined easily from the cascade of acyl intermediates produced by the Δ^{11}-desaturase and chain-shortening reactions (Figure 3).

SEX PHEROMONE PRODUCTION

Key Biosynthetic Enzymes

The overt chemical signals are only part of the chemistry involved in the overall communication system. The production of each pheromone component is dependent on specifically evolved chemical reactions taking place in special chemical factories (secretory cells) in the producing insect. Key to the biosynthesis are enzymes that either convert exogenous materials (such as oleic acid) into pheromone precursors or produce unique compounds from mevalonic acid or acetate. As shown above with the cabbage looper moth, a unique Δ^{11}-desaturase is found in lepidopteran sex pheromone glands to produce a whole variety of precursors in combination with chain-shortening reactions, followed by the action of reductases and acetyltransferases to produce the acetate pheromone components.

The enzymes themselves are only part of the chemistry involved at this stage. A more complete understanding of the biosynthesis includes studies on the structure of the enzymes, their cofactors, and the corresponding mRNA and gene. By using the Δ^{11}-desaturase as an example, it probably functions similarly to the stearoyl-CoA desaturase in animals, which has been found to be the terminal component of the electron transport system of endoplasmic reticulum that utilizes cytoplasmic NADH and molecular oxygen to effect Δ^9-desaturation of CoA esters of long-chain fatty acids (13). Research with liver microsomes has shown that there are three enzyme components of this sequence: a NADH-cytochrome b_5 reductase (a flavoprotein), cytochrome b_5 (a hemoprotein), and the Δ^9-desaturase (cyanide-sensitive factor). All three enzyme components are oriented with their catalytic domains on the cytoplasmic side of the endoplasmic reticulum, but since the cytochrome enzymes are solubilized more readily than is the desaturase, it was suggested that the desaturase protein is essentially buried in the microsomal membrane with just a small portion of it containing the active center exposed to the cytoplasm.

Stearoyl-CoA desaturase from rat liver has been purified and identified to be a protein with 358 amino acids and a molecular weight of 41,400 (14). The mRNA for this desaturase was found to be 4900 bases long, although a 1074-base open reading frame was found to code for a protein consisting of 358 amino acids of which 62% are hydrophobic. Two stearoyl-CoA desaturases have been characterized from mouse adipose tissue (15, 16) with 92 and 86% homology to that of the rat, and one from yeast (17) shows 36% identity and 60% similarity to the rat desaturase. The rat Δ^9-desaturase is able to functionally replace the yeast Δ^9-desaturase. Research on the characterization of the unique Δ^{11}-desaturase in moth pheromone glands is based on the tenet that the specialized desaturases in the pheromone glands are structurally and functionally related to the common Δ^9-desaturase that performs a general function in cellular lipid metabolism of animals and fungi. It must be determined whether the two desaturases are derived from independent transcription units or whether they are produced from the same transcription unit by a developmentally regulated alternative splicing mechanism. Data on the Δ^{11}-desaturase used in the biosynthesis of sex pheromone components will be important in addressing questions on how this enzyme evolved and what role this process played in speciation.

Hormonal Control

Another aspect of the sex pheromone communication system concerns the endogenous signals that control pheromone production and release from the emitting insect. A number of hormones have been found to be involved in the control of pheromone production in various insect species (18). Juvenile hormone was found to induce vitellogenesis and sex pheromone production in some cockroach and beetle species. However, ecdysteroids were found to be involved in regulating reproductive processes, including vitellogenin synthesis, in dipteran species.

In moths, it was discovered in *Helicoverpa zea* that a peptide produced in the subesophageal ganglion portion of the brain complex regulates pheromone production in female moths (19). This factor has been purified and characterized in three species, *Helicoverpa zea* (20), *Bombyx mori* (21, 22), and *Lymantria dispar* (23). They are all a 33- or 34-amino acid peptide (named pheromone biosynthesis activating neuropeptide, PBAN) and have in common an amidated C-terminal 5-amino acid sequence (FXPRL-amide), which is the minimum peptide fragment required for pheromontropic activity. In the redbanded leafroller moth, it was shown that PBAN from the brain stimulates the release of a different peptide from the bursae copulatrix that is used to stimulate pheromone production in the pheromone gland found at the posterior tip of the abdomen (24).

The chemistry involved in the mode of action of PBAN is complex and still under investigation, but present studies have shown that PBAN can interact directly with the pheromone gland and apparently stimulates pheromone biosynthesis by activating a plasma membrane calcium channel that regulates production of cAMP (25) (Figure 4). Other enzymes and effectors are likely involved in the signal transduction but remain uncharacterized. Several studies have been carried out to determine which enzyme(s) in the pheromone biosynthetic pathway is (are) controlled by PBAN. It apparently affects a step in or prior to fatty acid synthesis in several species (26, 27), perhaps involving acetyl-CoA carboxylase, although reduction of fatty acids to alcohols (reductase) was found to be controlled by PBAN in some other species (28, 29).

Studies of PBAN at the molecular level have shown that PBAN is synthesized as part of a larger precursor from which it is derived by posttranslational processing (30, 31). The two PBAN genes identified to date encode not only the PBAN peptide but also four additional neu-

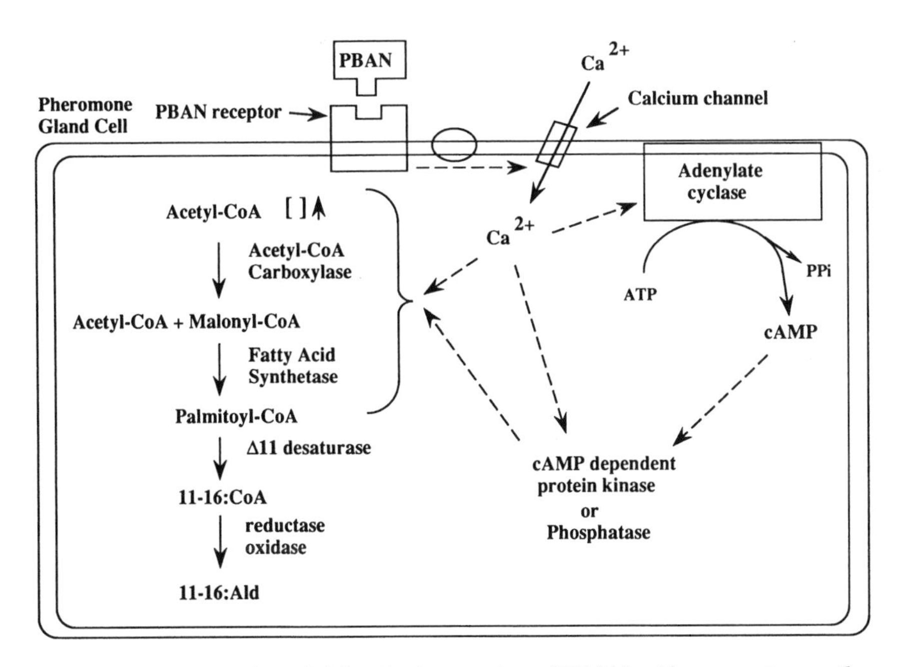

FIGURE 4 Proposed model for the interaction of PBAN with a receptor on the sex pheromone gland of a corn earworm female, and the resulting stimulation of acetyl-CoA carboxylase of the pheromone biosynthetic pathway.

ropeptides having the common C-terminal pentapeptide motif (FXPRL-amide) that is needed for pheromonotropic activity and is common with insect pyrokinin and myotropin peptide families. It is not clear if all five peptides are involved in the control of pheromone production, but one of them has been identified as the egg diapause hormone in *Bombyx mori* (32).

One additional factor that comes into play in the overall chemistry of the communication system relates to chemical signals from host plants that can override the photoperiodic control of phermone production. With the corn earworm, it was found that a volatile chemical signal from corn silk, probably ethylene, was required by wild insects for stimulation of pheromone production (33). This signal probably interacts with controls on the photoperiodic release of PBAN.

PHEROMONE PERCEPTION

Peripheral Detection

Insects can detect just a few hundred molecules of their sex pheromone in the environment with their highly efficient antennae, which can contain several hundred thousand receptor cells. A more detailed description of this highly efficient olfactory system will be presented by Hildebrand (34). The chemistry of perception, however, is just starting to be unraveled and still presents some interesting challenges. At the antennal level, this includes pheromone binding, transport to specific receptor sites, catabolism, and the process of signal transduction.

Chemistry responsible for the great specificity of insect pheromone receptors for single compounds is still unknown, but research has shown the involvement of special binding proteins in the receiving insect's antennae for pheromone transportation and sensillar enzymes for catabolism (35). Evidence also is accumulating that the chemical signals are transduced into intracellular second messenger responses via G-protein-coupled reaction cascades (36), involving phospholipase C, inositol trisphosphate, and the calcium channel. Additional data suggests that the cGMP cascade is involved secondarily in adaptation and fine tuning of the sensitivity of receptor cells (37, 38).

Plume Structure

The chemistry of signal transduction is important in the overall communication system, and a key factor in this process relative to obtaining an appropriate behavioral response is the frequency at which the pheromone components interact with the receptor cells. This frequency depends to a great extent on the release rate of the signal chemicals and the plume structure that carries them to the receiving insect. With moths, it has been found that the pheromone chemicals do not actually attract a mate by means of a chemical gradient, but rather the correct blend and abundance of pheromone molecules in a fluctuating plume will turn on two programs, optomotor anemotaxis and self-steered counterturning, in flying male moths and mediate flight upwind to the odor source (39).

Detection of pheromone, thus, turns on a visually guided motor program that produces upwind surges by the male moth, and when the signal is lost, the male stops upwind flight and the counterturning program causes it to cast sideways back and forth until the chemical signals are once again detected. The fluctuating plume structure is an integral part of the signal since the antennal receptors need intermittent stimulation to sustain upwind flight. The male moth will not fly upwind with continuous stimulation from a uniform plume, and high concentrations of pheromone that bombard the antennal receptors with plume filaments at a high frequency will cause the male to undergo arrestment of upwind flight. Thus, the patterning of the pheromone is as important as the composition.

Behavioral Thresholds

The processing of pheromone quality continues in the central nervous system (see ref. 34), involving the integration of odor input with visual stimuli and finally with messages sent to motor neurons for the appropriate behavioral responses. However, the chemistry involved in the brain is more complex than the processing of signals. Typically, the receiving insect becomes "receptive" to the sex pheromones at a specific time when the emitting insect is "calling." Although photoperiod cues and temperature are important in the diel periodicity of this response, it has been shown that biogenic amines also are involved (40).

Initially, it was found that injections of octopamine greatly enhanced a male's sensitivity to the chemical signal and that serotonin expanded the period of time in the scotophase during which males responded to pheromone. Recently, studies involving analyses of biogenic amines in various brain tissues showed that there was a significant correlation between decreased levels of octopamine in the brain in scotophase and the probability of a male to respond to low doses of pheromone in the flight tunnel (41). Decreased levels of octopamine also were correlated with increased levels of *N*-acetyloctopamine, suggesting that octopamine was being utilized in these tissues during that time period. Thus, processing and integration of the chemical signal in the central nervous system involves not only the neural pathways but also the endogenous influence of octopamine as it exerts a modulatory effect on the male's sensitivity to the signal.

Brain Black Box

Beyond the chemistry of the communication system, as described above, is an unknown and speculative factor that operates in the brain of male moths. It is a factor that seems to play a role in determining what is an appropriate signal for that male and what is not. The primary evidence for this factor comes from research with the European corn borer moth. Data show that a sex-linked gene(s) determines whether the male responds to the 97:3 *Z*/*E* pheromone blend of one race or to the 1:99 *Z*/*E* pheromone blend of another race (42, 43). The product of this gene(s) appears to function in the central nervous system and is not linked with the autosomal factor whose product effects different electrophysiological responses to pheromone components in antennal olfactory cells or with the autosomal factor controlling pheromone blend composition. These genetic studies involving various crosses and backcrosses of *E*- and *Z*-pheromone parents show that male moths will respond behaviorally as determined by the sex-linked factor, regardless of what antennal type they possess.

Additionally, an undescribed brain factor could be involved in the evolution of pheromone blends as input from particular pheromone components is switched from being "good" to one that elicits an antagonistic response from the male. Some evidence for this is when highly specific antennal receptors are found in the male's olfactory hairs that respond to a behaviorally antagonistic compound, and the special biosynthetic precursors for this component still are found in the female sex pheromone gland as evidence that at one time the component could have been part of the pheromone blend. An example is with the European corn borer. The male antennae have olfactory cells for the *E*11- and *Z*11–14:OAc pheromone components, but also exhibit one specialized cell that responds only to *Z*9–14:OAc—a compound that is antagonistic to the behavioral responses to pheromone (44). Also, the corn borer female pheromone gland contains large amounts of *Z*11–16:acid, which is a precursor for *Z*9–14:OAc, a compound that apparently no longer is produced or has a function in the pheromone system.

Many pheromone systems are finely tuned to a specific blend of components. Although some of the specificity can be effected by blend-specific sensory cells and other attributes of the neural system, there are still many questions related to the perception of pheromone blends. Included are questions not only on how the male changes its response to pheromone components produced by the female over time and how antagonists evolved but also on how a male can respond to a wider range of blend ratios under different environmental conditions (e.g., higher temperatures), how a male handles redundant pheromone signals, and how males might have changed their behavioral thresholds to higher or lower ones over time and in allopatry.

The chemistry of sex attraction, thus, presents challenges not only among the diverse animal species but also within each communication system. Some of the chemistry has been defined, as described above, but knowledge of the chemistry in a few insect species pales in view of the vast number of frontiers yet to be attacked in other animal species. Similarities will, no doubt, be found in the communication systems of other animals, but the diversity of nature predicts that there are still many surprises to be discovered by tomorrow's chemists.

SUMMARY

The chemical communication system used to attract mates involves not only the overt chemical signals but also indirectly a great deal of chemistry in the emitter and receiver. As an example, in emitting female moths, this includes enzymes (and cofactors, mRNA, genes) of the pheromone biosynthetic pathways, hormones (and genes) involved in controlling pheromone production, receptors and second messengers for the hormones, and host plant cues that control release of the hormone. In receiving male moths, this includes the chemistry of pheromone transportation in antennal olfactory hairs (binding proteins and sensillar esterases) and the chemistry of signal transduction, which includes specific dendritic pheromone receptors and a rapid inositol triphosphate second messenger signal. A fluctuating plume structure is an integral part of the signal since the antennal receptors need intermittent stimulation to sustain upwind flight. Input from the hundreds of thousands of sensory cells is processed and integrated with other modalities in the central nervous system, but many unknown factors modulate the information before it is fed to motor neurons for behavioral responses. An unknown brain control center for pheromone perception is discussed relative to data from behavioral-threshold studies showing modulation by biogenic amines, such as octopamine and serotonin, from genetic studies on pheromone discrimination, and from behavioral and electrophysiological studies with behavioral antagonists.

I thank Drs. Russell Jurenka and Charles Linn for their assistance in preparing this manuscript and for their role, along with many other excellent scientists in my laboratory, for conducting much of the research discussed in this paper. Recent research has been sponsored by National Science Foundation Grants IBN-9108743 and IBN-9017793, National Institutes of Health Grant AI-32498, and a grant from the Cornell Center for Advanced Technology in Biotechnology.

REFERENCES

1. Lintner, J. A. (1882) *West. N. Y. Hortic. Soc. Proc.* **27,** 52–66.
2. Mayer, M. S. & McLaughlin, J. R. (1991) *Handbook of Insect Pheromones and Sex Attractants* (CRC, Boca Raton, FL).
3. Butenandt, A., Beckman, R., Stamm, D. & Hecker, E. (1959) *Z. Naturforsch. B* **14,** 283–284.
4. Schneider, D. (1969) *Science* **163,** 1031–1037.
5. Schneider, D. (1992) *Naturwissenschaften* **79,** 241–250.
6. Roelofs, W. L. (1984) in *Techniques in Pheromone Research,* eds. Hummel, H. E. & Miller, T. A. (Springer, New York), pp. 131–159.
7. Charlton, R. E., Webster, F. X., Zhang, A., Schal, C., Liang, D., Sreng, I. & Roelofs, W. L. (1993) *Proc. Natl. Acad. Sci. USA* **90,** 10202–10205.
8. Roelofs, W. L., Comeau, A., Hill, A. & Milicevic, G. (1971) *Science* **174,** 297–299.
9. Persoons, C. J., Verwiel, P. E. J., Ritter, F. J., Talman, E., Nooyen, P. E. J. & Nooen, W. J. (1976) *Tetrahedron Lett.* **24,** 2055–2058.
10. Silverstein, R. M. (1979) in *Chemical Ecology: Odour Communication in Animals,* ed. Ritter, F. J. (Elsevier/North–Holland, Amsterdam), pp. 133–146.
11. Lanier, G. N., Classon, A., Stewart, T., Piston, J. J. & Silverstein, R. M. (1980) *J. Chem. Ecol.* **6,** 677–687.
12. Bjostad, L. B., Wolf, W. A. & Roelofs, W. L. (1987) in *Pheromone Biochemistry,* eds. Prestwich, G. D. & Blomquist, G. J. (Academic, New York), pp. 77–120.

13. Jeffcoat, R. (1979) *Essays Biochem.* **15,** 1–36.
14. Thiede, M. A., Ozols, J. & Strittmatter, P. (1986) *J. Biol. Chem.* **261,** 13230–13235.
15. Ntambi, J. M., Buhrow, S. A., Kaestner, K. H., Christy, R. J., Sibley, E., Kelly, T. J., Jr., & Lane, M. D. (1988) *J. Biol. Chem.* **263,** 17291–17300.
16. Kaestner, K. H., Ntambi, J. M., Kelly, J., T. J. & Lane, M. D. (1989) *J. Biol. Chem.* **264,** 14755–14761.
17. Stukey, J. E., McDonough, V. M. & Martin, C. E. (1989) *J. Biol. Chem.* **264,** 16537–16544.
18. Prestwich, G. D. & Blomquist, G. J. (1987) *Pheromone Biochemistry* (Academic, New York).
19. Raina, A. K. & Klun, J. A. (1984) *Science* **225,** 531–533.
20. Raina, A. K., Jaffe, H., Kempe, T. G., Keim, P., Blacher, R. W., Fales, H. M., Riley, C. T., Klun, J. A., Ridgway, R. L. & Hayes, D. K. (1989) *Science* **244,** 796–798.
21. Kitamura, A., Nagasawa, H., Kataoka, H., Inoue, T., Matsumoto, S., Ando, T. & Suzuki, A. (1989) *Biochem. Biophys. Res. Commun.* **163,** 520–526.
22. Kitamura, A., Nagasawa, H., Kataoka, H., Ando, T. & Suzuki, A. (1990) *Agric. Biol. Chem. Tokyo* **54,** 2495–2497.
23. Masler, E. P., Raina, A. K., Wagner, R. M. & Kochansky, J. P. (1994) *Insect Biochem. Mol. Biol.* **24,** 829–836.
24. Jurenka, R. A., Fabriás, G. & Roelofs, W. L. (1991) *Insect Biochem.* **21,** 81–89.
25. Jurenka, R. A. & Roelofs, W. L. (1993) in *Insect Lipids: Chemistry, Biochemistry and Biology*, eds. Stanley-Samuelson, D. W. & Nelson, D. R. (Univ. of Nebraska Press, Lincoln), pp. 353–388.
26. Jurenka, R. A., Jacquin, E. & Roelofs, W. L. (1991) *Arch. Insect Biochem. Physiol.* **17,** 81–91.
27. Jacquin, E., Jurenka, R. A., Ljungberg, H., Nagnan, P., Löfstedt, C., Descoins, C. & Roelofs, W. L. (1993) *Insect Biochem. Mol. Bol.* **24,** 203–211.
28. Martinez, T., Fabriás, G. & Camps, F. (1990) *J. Biol. Chem.* **265,** 1381–1387.
29. Arima, R., Takahara, K., Kadoshima, T., Nagasawa, H., Kitamura, A. & Suzuki, A. (1991) *Appl. Ent. Zool.* **26,** 137–147.
30. Kawano, T., Kataoka, H., Nagasawa, H., Isogai, A. & Suzuki, A. (1992) *Biochem. Biophys. Res. Commun.* **189,** 221–226.
31. Ma, P. W. K., Knipple, D. C. & Roelofs, W. L. (1994) *Proc. Natl. Acad. Sci. USA* **91,** 6506–6510.
32. Sato, Y., Nakazawa, Y., Menjo, N., Imai, K., Komiya, T., Saito, H., Shin, M., Ikeda, M., Sakakibara, K., Isobe, M. & Yamashita, O. (1992) *Proc. Jpn. Acad. Ser. B* **68,** 75–79.
33. Raina, A. K., Kingan, T. G. & Mattoo, A. K. (1992) *Science* **255,** 592–594.
34. Hildebrand, J. G. (1995) *Proc. Natl. Acad. Sci. USA* **92,** 67–74.
35. Vogt, R. G. (1991) in *Pheromone Biochemistry*, eds. Prestwich, G. D. & Blomquist, G. J. (Academic, New York), pp. 385–431.
36. Stengl, M., Hatt, H. & Breer, H. (1992) *Annu. Rev. Physiol.* **54,** 665–681.
37. Boekhoff, I., Seifert, E., Göggerle, S., Lindemann, M., Krüger, B.-W. & Breer, H. (1993) *Insect Biochem. Mol. Biol.* **23,** 757–762.
38. Kaissling, K.-E. & Boekhoff, I. (1993) in *Sensory Systems of Arthropods*, eds. Wiese, K., Gribakin, F. G., Popov, A. V. & Renninger, G. (Birkhaeuser, Basel), pp. 489–502.
39. Vickers, N. J. & Baker, T. C. (1994) *Proc. Natl. Acad. Sci. USA* **91,** 5756–5760.
40. Roelofs, W. L. & Linn, C. E., Jr. (1987) *J. Cell Biochem. Suppl.* **10,** 54.

41. Linn, C. E., Jr., Campbell, M. G., Poole, K. R. & Roelofs, W. L. (1994) *Comp. Biochem. Physiol.* **1086,** 87–98.
42. Roelofs, W. L., Glover, T. J., Tang, X.-H., Sreng, I., Robbins, P., Eckenrode, C. J., Lofstedt, C., Hansson, B. S. & Bengtsson, B. O. (1987) *Proc. Natl. Acad. Sci. USA* **84,** 7585–7589.
43. Glover, T. J., Campbell, M. G., Robbins, P. S. & Roelofs, W. L. (1990) *Arch. Insect Biochem. Physiol.* **15,** 67–77.
44. Glover, T. J., Perez, N. & Roelofs, W. L. (1989) *J. Chem. Ecol.* **15,** 863–873.

The Chemistry of Sexual Selection

THOMAS EISNER AND JERROLD MEINWALD

Chemical dependencies are fundamental in nature. Animals as a group depend on plants for that most basic of metabolites, glucose, which plants produce through photosynthesis. Countless organisms, ourselves included, have vitamin and other dietary requirements. Insects synthesize steroidal hormones but can produce these only from other steroids, such as cholesterol, which they need to obtain with the diet (1). In recent years there has been considerable interest in animals, mostly insects, that depend on exogenous compounds for defense (2). The strategy has human parallels. Many of the medicinals we use against parasites and pathogens are obtained from nature. In insects, the acquired compounds protect primarily against predators. The monarch butterfly, for instance, is distasteful to birds by virtue of cardenolides that it sequesters from its larval foodplants (3).

We here summarize work we have done with a moth, *Utetheisa ornatrix*, that has a dependence on certain plant alkaloids. The moth uses the compounds for defense and for production of a pheromone that plays a decisive role in sexual selection. The species has a broad range, extending through North America east of the Rockies and southward into Brazil, Argentina, and Chile. Our studies were done mostly with populations of the moth from central Florida.

Thomas Eisner is Schurman Professor of Biology and director of the Cornell Institute for Research in Chemical Ecology and Jerrold Meinwald is Goldwin Smith Professor of Chemistry at Cornell University, Ithaca, New York.

DEFENSE

Utetheisa, like many other Lepidoptera of the family Arctiidae, is aposematic. White, with pink hindwings and black and yellow markings (Figure 1*A*), it is highly conspicuous on the wing. It flies as readily in daytime as at night. We suspected the moth to be unpalatable, and we were able to prove this in experiments with orb-weaving spiders.

We knew from previous work that moths are protected from entanglement in spider webs by their investiture of scales. Instead of sticking to webs as "naked" insects typically do, they simply lose scales to points of contact with the orb and flutter loose (4). *Utetheisa*, in contrast, becomes instantly quiescent when it flies into a web. The spider converges on the moth and inspects it, but then, almost invariably, sets it free. During the inspection the moth sometimes emits its defensive froth (Figure 1*C*), but it is released even if it withholds the effluent. Spiders such as *Nephila clavipes* cut the moth from the web by snipping the entangling threads with their fangs (5, 6) (Figure 2*A*), while others, such as *Argiope florida*, free the moth by pulling it from the web (T.E., unpublished observations).

We knew *Utetheisa* to feed on poisonous plants as a larva (Figure 1*B*). The plants, of the genus *Crotalaria* (family Leguminosae), were known to contain pyrrolizidine alkaloids (henceforth abbreviated as PAs), intensely bitter compounds potently hepatotoxic to mammals (7). Other species of *Utetheisa* were known to sequester PAs (8). We found this to be true for *U. ornatrix* as well. Adult Utetheisa raised on *Crotalaria spectabilis*, one of the principal foodplants available to the moth in the United States, contain on average about 700 μg of monocrotaline (**1**), the principal PA in that plant (9, 10).

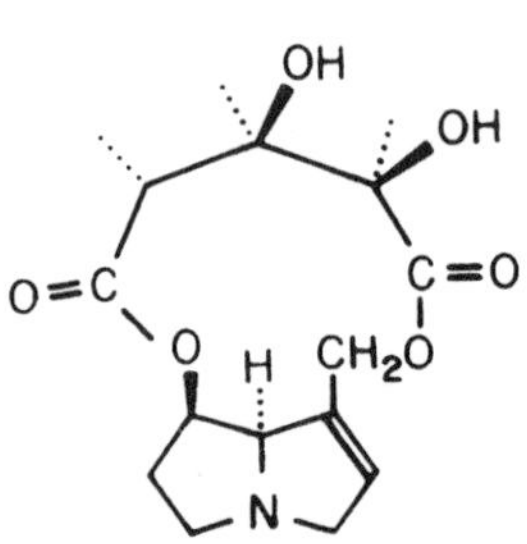

I

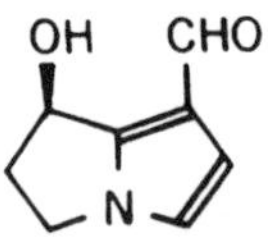

II

In the laboratory, we succeeded in raising *Utetheisa* on two alternative artificial diets, one made up with *Crotalaria* seeds and containing the PA monocrotaline (CS diet), the other based on pinto beans and devoid of PA (PB diet). On the assumption that the PB diet-raised moths, which we proved to be PA-free, would be vulnerable to predation, we took moths

FIGURE 1 Photos of *Utetheisa ornatrix*. (*A*) Adult, at rest, on pods of one of the larval foodplants (*Crotalaria mucronata*). (*B*) Larva within pod of another of the foodplants (*Crotalaria spectabilis*). (*C*) Adult, emitting defensive froth. (*Continued overleaf.*)

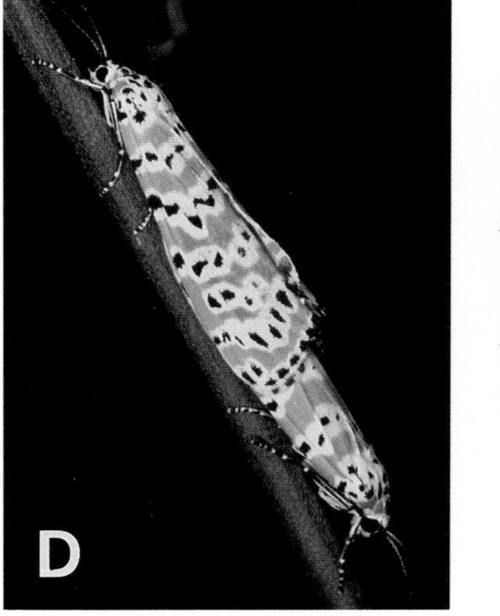

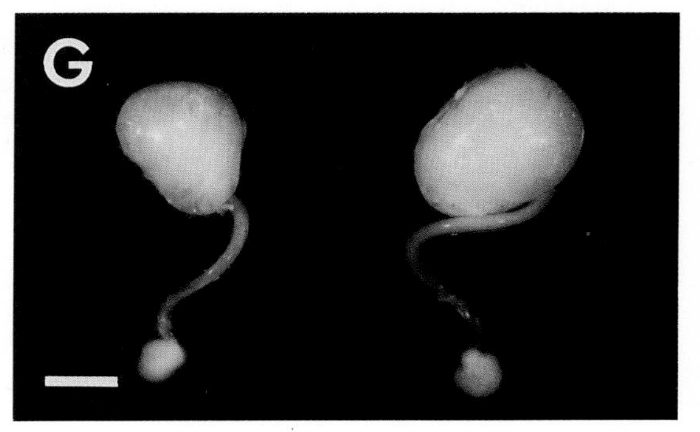

FIGURE 1 (*Continued*)

(*D*) Adults mating. (*E*) Adult male, with coremata everted, courting female (laboratory test). (*F*) Coremata, everted. (*G*) Spermatophores, excised from bursa of just-mated females. (Bar in *A* = 1 cm; bars in *F* and *G* = 1 mm.)

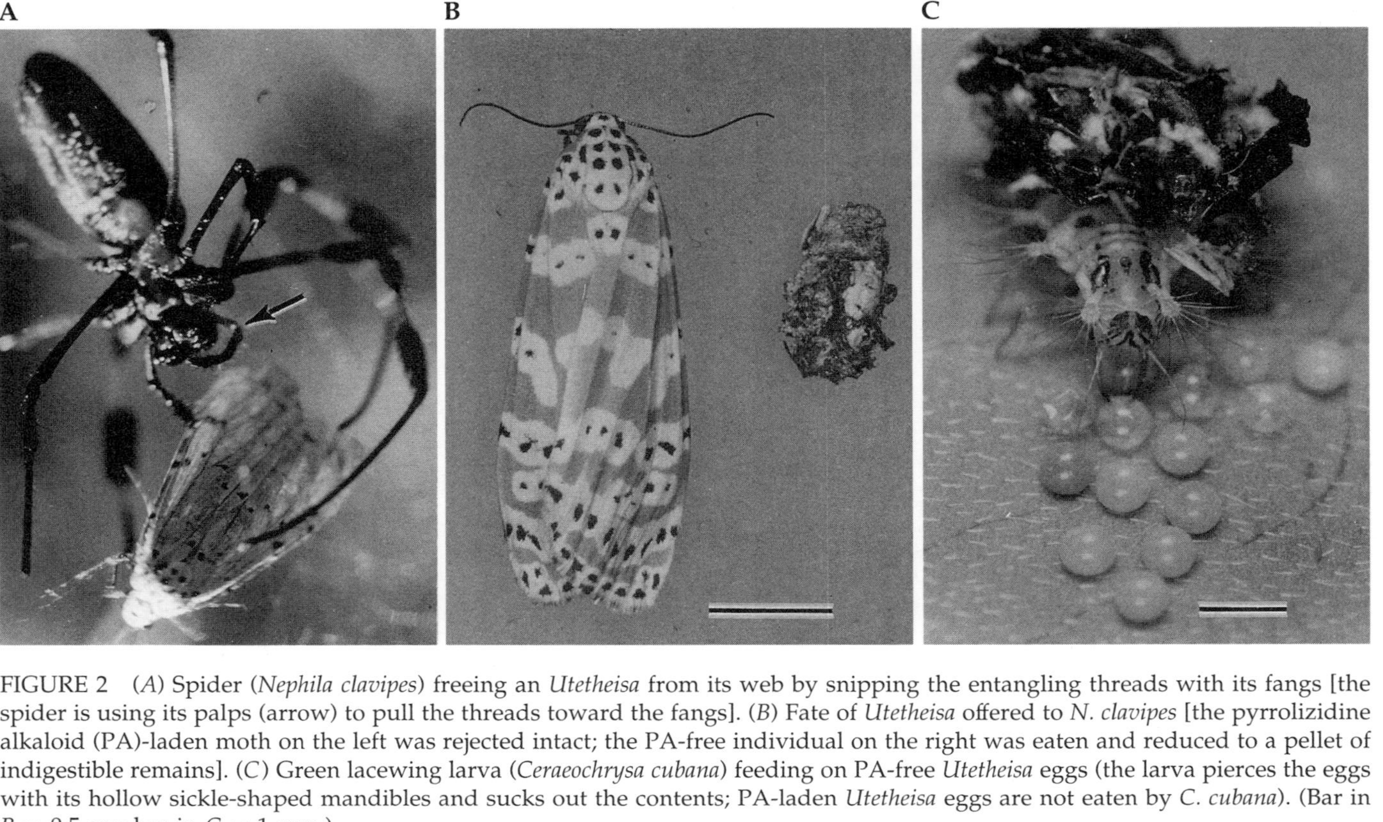

FIGURE 2 (*A*) Spider (*Nephila clavipes*) freeing an *Utetheisa* from its web by snipping the entangling threads with its fangs [the spider is using its palps (arrow) to pull the threads toward the fangs]. (*B*) Fate of *Utetheisa* offered to *N. clavipes* [the pyrrolizidine alkaloid (PA)-laden moth on the left was rejected intact; the PA-free individual on the right was eaten and reduced to a pellet of indigestible remains]. (*C*) Green lacewing larva (*Ceraeochrysa cubana*) feeding on PA-free *Utetheisa* eggs (the larva pierces the eggs with its hollow sickle-shaped mandibles and sucks out the contents; PA-laden *Utetheisa* eggs are not eaten by *C. cubana*). (Bar in *B* = 0.5 cm; bar in *C* = 1 mm.)

from both cultures into the field and offered them to *N. clavipes.* The spiders consistently freed the moths raised on CS diet [whose PA content matched that of field-raised *Utetheisa* (10)] but killed and consumed the PA-free controls (Figure 2*B*). We also tested directly for the deterrency of PA. We added crystalline monocrotaline to edible items (mealworms) ordinarily consumed by *N. clavipes* and found that by doing so we could render such items relatively unacceptable to the spiders (6).

Adult *Utetheisa* tend to be rejected also by birds (blue jays, *Cyanocitta cristata*; scrub jays, *Aphelocoma coerulescens*; T.E., unpublished observations), as might be expected, given their aposematism, but there is no definitive evidence that the unacceptability is due specifically to the PAs.

Tests with larval *Utetheisa* showed these to be rejected by wolf spiders, but only if the larvae had fed on *Crotalaria* or CS diet. Larvae raised on PB diet proved consistently palatable to the spiders (11).

COURTSHIP

Utetheisa, like many other insects, court at dusk (Figure 1*D*). The female initiates the behavior. Positioned on vegetation (often on branches of *Crotalaria* itself), she emits a sex attractant that drifts downwind and lures males (12). Caged virgin females, placed outdoors, attract males (Figure 3*A*). The glands that produce the pheromone are a pair of long, coiled tubes, opening close together near the abdominal tip (Figure 3*C*). Extraction of the glands led to the characterization of a long-chain polyene (**3**), which proved electrophysiologically active on male antennae

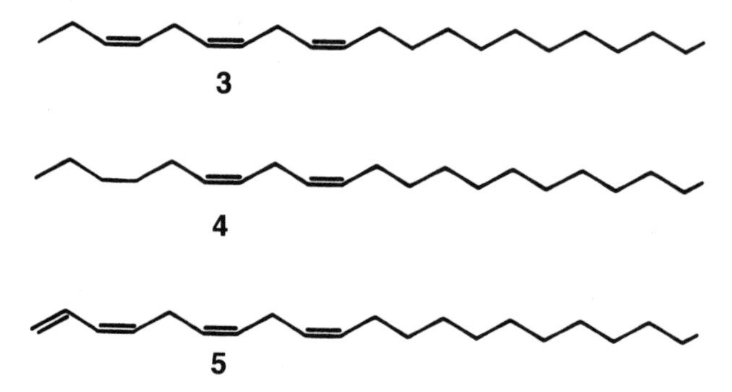

[electroantennogram (EAG) tests] and attractive to males in the field (12) (Figure 3*B*). The pheromone has since been found to contain two additional polyenes (**4** and **5**) (13).

We found female *Utetheisa* to emit their attractant in pulses (1.5 ± 0.2 pulses per s) (12) (Figure 3*D*), as some other moths have since also been shown to do (14, 15). Pulsation frequencies are similar in the various species, indicating that there is no informational specificity associated with the temporal modulation of the signals. Views differ as to the function of pheromone pulsation in moths (12, 14–17). A reasonable suggestion, based on mathematical modeling, is that it increases the efficiency of signal output. By pulsing, females can presumably increase the range over which they are able to attract males from downwind (18).

SEXUAL SELECTION

Courtship in *Utetheisa* involves more than mere attraction of males by females. Once a male locates a female, the pair does not at once proceed to mate. The male first flutters around the female, hovering beside her or circling her at close range, while at the same time occasionally thrusting his abdomen against her. It is only after one or more such thrusts that the female parts her wings and presents her abdomen for mating. We analyzed this behavior, which takes place in darkness, from videotapes taken under infrared illumination (19). We noted that when the male executes his abdominal thrusts, he everts from the abdominal tip a pair of brushes that he ordinarily keeps tucked away in pouches (Figure 1 *E* and *F*). The brushes, called coremata, are secretory. Each consists of a tuft of modified scales, associated with glandular tissue at the base, and wetted by secretion. Extraction of the brushes revealed presence of a compound, hydroxydanaidal (HD; **2**) (19), previously found in coremata of other *Utetheisa* species (20). The structure of HD suggested that the compound was derived from PA. We predicted that *Utetheisa* raised on PA-free diet should have HD-free coremata, which turned out to be the case (19).

We showed that corematal excision rendered males less acceptable to females (mock-operated males were not thus handicapped). This proved that the coremata had a function, but did not in itself provide evidence for the role of HD. However, we found that PB diet-raised males, whose coremata were normal except for lack of HD, were also relatively unsuccessful in courtship (19).

Further data showed HD to have a direct stimulatory effect on the female. When quiescent females were stroked with excised coremata, they tended to present their abdomen, but only if the coremata contained HD. Moreover, to elicit maximal effect, the HD had to be of the absolute configuration [*R*(−)] in which the compound occurs in the coremata (19). Predictably, female *Utetheisa* were found to have antennal chemoreceptors highly sensitive to the *R*(−) isomer of HD (21).

We wondered about the nature of the corematal message. Was HD simply the male's way of announcing his presence to the female, or was the molecule conveying more subtle information? Specifically, we asked whether the derivation of a pheromone from phytotoxin might have

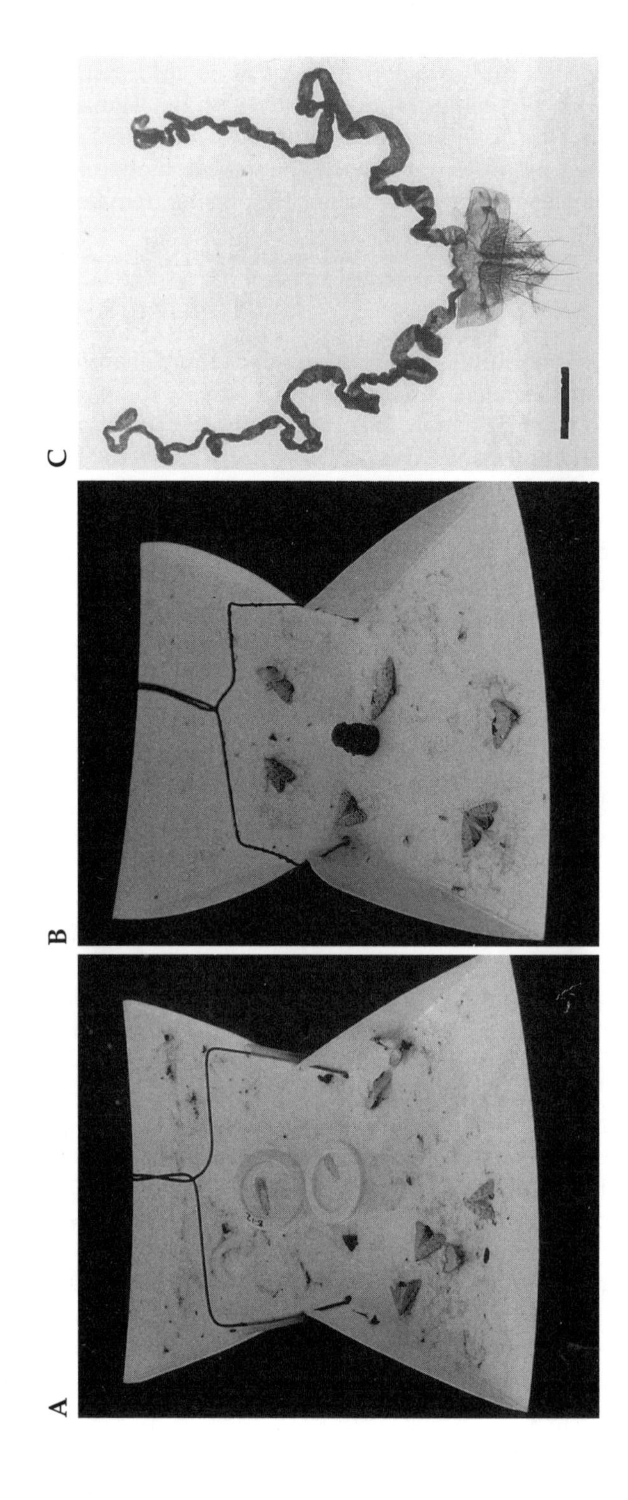
A
B
C

D

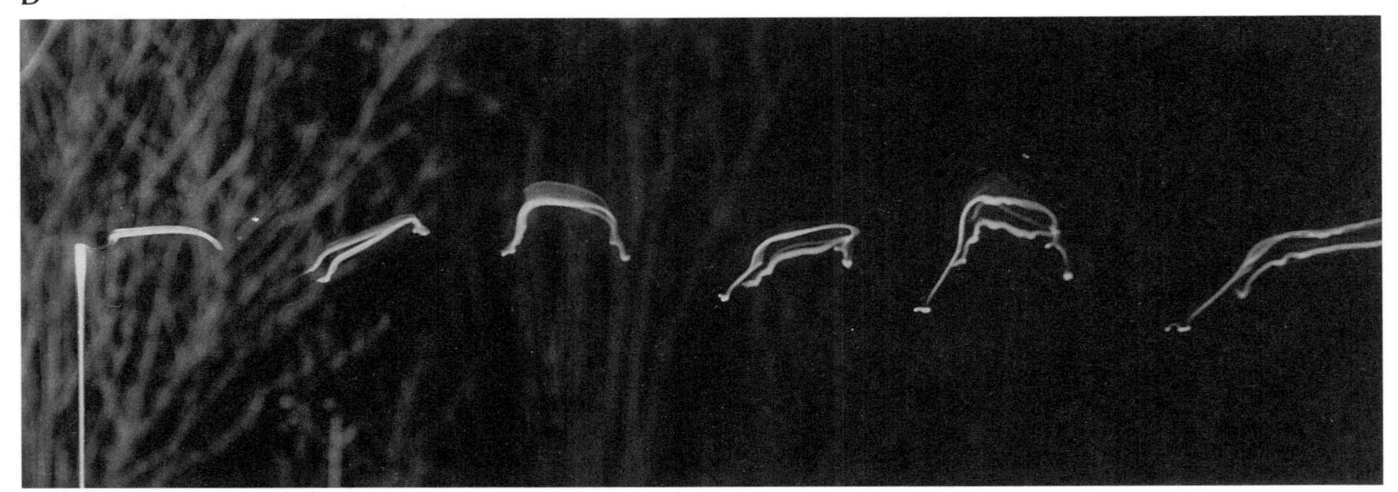

FIGURE 3 Male attraction in *Utetheisa*. (*A*) Virgin females, confined to small screened cages affixed to a tray placed outdoors, have attracted a number of males at dusk (the males have become trapped in the sticky coating of the tray). (*B*) Test comparable to the preceding one, but using synthetic compound **3** of the female's pheromone in lieu of females (the compound is in rubber cup in center of tray). (*C*) Glandular source of female pheromone (the glands, ordinarily more tightly coiled, are in posterior part of the female's abdomen). (*D*) Simulation of pulse emission pattern of female pheromone [a fluidics system has been used to generate a pulsed, visibly marked (titanium tetrachloride), plume of air]. (Bar in *C* = 0.5 mm.)

special adaptive significance. We postulated that HD could provide the male with a means for proclaiming his PA content, a parameter that could be variable, and which could provide the female with a basis for exercising mate choice (19, 22).

We knew that larval *Utetheisa*, in their later instars, feed predominantly on the seeds of *Crotalaria*, the parts of the plant richest in PA. We knew further that PA was a strong phagostimulant that drove larvae in their quest for food (10, 23). Moreover, field observation had told us that the seeds of *Crotalaria* were a variable resource, for which the larvae might need to compete at times. We also knew that adult *Utetheisa* differed substantially in their PA load in nature and that under experimental conditions their PA load varied in proportion to the seed content of the larval diet (9, 24). It remained to be seen whether the male's HD content was an indicator of his PA content. Chemical analyses proved that it was (24). Our postulate that HD could provide the female with a means of assessing the male's PA load, as well as possibly his larval competitive ability, a potentially heritable trait, was strengthened. But, as we were to learn, we were not being imaginative enough.

PARENTAL INVESTMENT

When we analyzed *Utetheisa* eggs, we discovered that they too contain PA, suggesting that they were being endowed, for their chemical protection, by the mother. However, we determined that eggs could contain PA even when the mother was herself PA-free. All that was necessary was for the father to be PA-laden. These results, together with others, established that the male transfers PA to the female at mating, and that both parents bestow PA upon the eggs (25). The allocation is uneven. By mating males and females laden with different PAs (usaramine and monocrotaline, respectively) and analyzing the eggs for both PAs, we were able to determine that the eggs receive only about a third of their PA from the father (25). However, these data pertained to once-mated females, when in fact female *Utetheisa* mate with a number of males over their life-span (mean number = 4–5 males; more than 10 partners have been recorded) (26). On average, therefore, eggs might receive more than a third of their PA from males.

The eggs are effectively protected by their alkaloidal endowment. Both coccinellid beetles and ants are deterred by PA dosages commensurate with levels naturally prevailing in *Utetheisa* eggs (*Utetheisa* eggs contain an average 0.8 μg of PA). With both predators, PA proved most effective as the N-oxide, the predominant form in which PA occurs in *Utetheisa* (25, 27). Also deterred were larvae of green lacewings (*Ceraeochrysa cubana*), which rejected eggs from moths raised on CS diet while accepting those from individuals raised on PB diet (T.E., unpublished observations) (Figure 2C).

We were forced to review our interpretation of the corematal message. HD, it seemed, could serve not only for proclamation of alkaloid load and of a genetic capacity, but for advertisement of a nuptial gift. We postulated that the magnitude of the male's PA offering should be proportional to his PA load, and we found this to be the case (24).

We determined further that in nature male PA content is proportional to male mass (9, 26), indicating that by favoring males of high PA content females could be selecting also for males of large size. Larger males transfer larger spermatophores, and thus more nutrient for investment in eggs (with each mating beyond the first, the female is able to produce, on average, an extra 32 eggs, an equivalent of upward of 10% of her basic output) (26, 28). The advertisement implicit in the corematal signal could thus be for both PA and nutrient (26).

The correlation of large size and PA load in *Utetheisa* could have a simple reason. *Crotalaria* seeds are rich not only in PA but in nutrient. Larvae competing successfully for seeds could therefore inevitably be destined to achieve large adult size.

SPERM SELECTION

A question of interest concerned the paternity of *Utetheisa* offspring. Given that females are promiscuous, do males have assurance of fathering offspring when they mate? Is there sperm mixing in multiply mated females, or do the sperm of some males "win out" over those of others?

By means of enzymatic markers we checked into offspring paternity of twice-mated females. We found the progeny in most cases to be sired almost exclusively by the larger of the two males. Factors such as duration of copulation, mating order, or between-mating interval were not determinants of male "success." Nor was male PA content, which in our laboratory-raised *Utetheisa* did not correlate with male size (we attribute this lack of correlation to the fact that our artificial larval diets were equally nutritious, whether PA-laden or PA-free) (26).

We have evidence that the female herself controls the mechanism by which one set of sperm is favored over the other. If females are anesthetized so as to inactivate their muscles, including presumably the many pumping muscles of their reproductive organs, the normal routing of sperm is inhibited. The anesthesia does not immobilize the sperm, indicating that it is not the sperm themselves that are in control of their fate (26). The reproductive system of female *Utetheisa* is a complex labyrinth of ducts and chambers. While we do not know precisely how the various components of the system operate, we have hypotheses on how the female might selectively retain or expel sets of sperm.

We also have evidence of how the female assesses the size of her mating partners. She appears to do so indirectly, by gauging the size of their spermatophore, for which purpose she may use stretch receptors that female moths are known to have in the chamber (bursa) in which spermatophores are deposited (29). Male *Utetheisa* can be caused to produce inordinately small spermatophores if they are mated relatively recently beforehand. If such mated males are placed in competition with physically smaller males, whose spermatophores may now be the relatively larger ones, they tend to "lose out" (30). We predict from this that males, in nature, may space their matings days apart. To regain the capacity to produce full size spermatophores takes a male about a week (30).

The female strategy is an interesting one. By accepting multiple partners she can accrue multiple gifts, to her obvious benefit, since she can thereby promote both her fecundity and the survivorship of her offspring. By discriminating between sperm, she is able to select for traits that in the genetic sense have long-range payoff. By favoring sperm of large males the female is essentially reinforcing, *after* copulation, the choice mechanism that she already exercised in the precopulatory context. Postcopulatory assessment provides the female with the option of taking corrective action. If on a given evening she accepted a male of moderate size and PA content, she can still discriminate genetically against that male by utilizing the sperm of a larger, more PA-laden and therefore genetically superior individual, that she is able to lure on a subsequent night. But the earlier mating is canceled in a genetic sense only. Nutrient and PA that the female receives from the losing male are utilized by her, as are all gifts that she obtains from males (26, 28).

We feel that we may have a tentative answer to the question of why smaller males, of lesser PA content, appear not to "lie" in the context of courtship. Could such males not masquerade as "desirable" by producing exaggerated levels of HD? Perhaps their failure to do so is a reflection of the fact that they would be "found out" unless they also produced outsized spermatophores. Smaller males, even if able to convert extra PA into HD for inflation of their chemical message, may lack the extra nutrient needed for inflation of the spermatophore. By putting the male to the test by way of a second criterion after mating, the female has the means to check on liars.

ADDITIONAL FINDINGS

We discovered in the laboratory that *Utetheisa* larvae deficient in PA can make up their chemical shortfall by resorting to cannibalism (10, 23).

They attack both pupae and eggs, and they appear to be driven to cannibalism not so much by hunger as by the PA deficiency itself. Moreover, they target specifically eggs and pupae that are PA-laden rather than PA-free. Possession of high systemic PA loads could therefore, under some circumstances, be endangering to *Utetheisa*, rather than beneficial. We do not know whether in nature *Utetheisa* are ever seriously at risk from cannibalism, although it is interesting to note that *Utetheisa* pupate out of reach of larval attack, in secluded sites away from the foodplant (23). The danger may be real for eggs, however, which are laid on the leaves of *Crotalaria* and therefore exposed to larvae.

A further finding concerns the stereochemistry of HD and its derivation from PA. Both HD and the primary PAs (monocrotaline, usaramine) that we know to be available to *Utetheisa* in the field are of the same (7*R*) stereochemical configuration. It was therefore not surprising to find that *Utetheisa* is unable to convert a PA of opposite (7*S*) stereochemistry (heliotrine) into HD. However, we found another arctiid moth, the Asian species *Creatonotus transiens*, which also produces HD in its coremata, to be able to use 7*R* and 7*S* PAs interchangeably for HD production (31). We are tempted to conclude that *Creatonotus*, unlike *Utetheisa*, has dietary access to PAs of both stereochemical configurations in its environment.

THE BROADER PERSPECTIVE

Parental bestowment of defensive substances upon eggs may be more widespread in insects than generally suspected. Insect eggs, by virtue of immobility alone, are highly vulnerable, and it makes sense that they should be protected. *Utetheisa*'s strategy of utilizing substances of exogenous origin for egg defense is not without parallel. Nor is the strategy of paternal involvement in the provisioning process.

In *Apiomerus flaviventris*, an assassin bug (family Reduviidae), the female alone provisions the eggs. She procures a terpenoid resin from plants and applies this to the eggs, thereby protecting these from ants (32). Blister beetles (family Meloidae) protect their eggs with cantharidin. The compound is biosynthesized by the beetles, sometimes by the males alone, which transfer it to the females, for incorporation into the eggs (33). Cantharidin is also utilized by *Neopyrochroa flabellata*, a fire-colored beetle (family Pyrochroidae). In this insect, the cantharidin is procured by the male from an unknown exogenous source and is also transmitted by way of the female to the eggs. Interestingly, the courting male advertises his possession of cantharidin by exuding a small amount of the compound as secretion from a cephalic gland. The female feeds on this secretion prior to mating and rejects males unable to provide such proof of "worth" (31).

The closest parallel to the *Utetheisa* strategy is exhibited by danaine butterflies (family Nymphalidae). In one of these, the queen butterfly (*Danaus gilippus*), which we have studied in detail, the adult male visits plants that produce PA, and he ingests PA from these sources. He then transfers the PA to the female at mating, and she bestows virtually the entire gift upon the eggs. Remarkably, the male produces a pheromone, danaidone, which he derives from PA, and which he "airs" in the context of courtship by everting two glandular brushes that secrete the compound. Males deficient in danaidone tend to be rejected in courtship (6, 34). Work done on other danaines, as well as on the related ithomiines, suggests that comparable behavior may be widespread in these butterflies (6, 35–38).

Paternal provisioning may also involve bestowal of inorganic compounds upon the eggs. Many butterflies and moths drink extensively at water sources, a phenomenon known as "puddling." The behavior is sex-biased and involves for the most part males. It had long been suspected that the benefit from puddling is sodium sequestration (39), and this has now been demonstrated (40, 41). It is also becoming clear that the acquired sodium is transferred in part to the female at mating (40, 41), and that the female transmits the gift to the eggs (41). Whether prior to mating males advertise their ionic merits by proclaiming in some fashion that they are "worth their salt" remains unknown.

SUMMARY

The moth *Utetheisa ornatrix* (Lepidoptera: Arctiidae) is protected against predation by pyrrolizidine alkaloids that it sequesters as a larva from its foodplants. At mating, the male transfers alkaloid to the female with the spermatophore, a gift that the female supplements with alkaloid of her own and transmits to the eggs. Eggs are protected as a result. The male produces a pheromone, hydroxydanaidal, that he derives from the alkaloid and emits from a pair of extrusible brushes (coremata) during precopulatory interaction with the female. Males rendered experimentally alkaloid-free fail to produce the pheromone and are less successful in courtship. The male produces the pheromone in proportion both to his alkaloid load and to the amount of alkaloid he transfers to the female. The pheromone could thus serve as an indication of male "worth" and provide a basis for female choice. *Utetheisa* females are promiscuous and therefore are able to accrue multiple nuptial gifts (alkaloid and nutrient, both transmitted with the spermatophore). They use sperm selectively, favoring those of larger males. Larger males in nature are also richer in alkaloid. Females therefore reinforce after copulation the choice mechanism they already exercise during courtship.

We are greatly indebted to the many associates who collaborated in our research on *Utetheisa* and to the staff of the Archbold Biological Station, Lake Placid, FL, where much of our field work was done. The study was supported largely by National Institutes of Allergy and Infectious Diseases Grants AI02908 and AI12020, by Hatch Grants NYC-191424 and NYC-191425, and by unrestricted funds from the Schering–Plough Research Institute and the Merck Research Laboratories.

REFERENCES

1. Clayton, R. B. (1964) *J. Lipid Res.* **5,** 3–19.
2. Rosenthal, G. A. & Berenbaum, M. R. (1992) *Herbivores: Their Interactions with Secondary Plant Metabolites* (Academic, New York), Vols. 1 and 2.
3. Brower, L. P., Nelson, C. J., Fink, L. S., Seiber, J. N. & Calhoun, B. (1988) in *Chemical Mediation of Coevolution*, ed. Spencer, K. C. (Academic, New York), pp. 447–475.
4. Eisner, T., Alsop, R. & Ettershank, G. (1964) *Science* **146,** 1058–1061.
5. Eisner, T. (1982) *BioScience* **32,** 321–326.
6. Eisner, T. & Meinwald, J. (1987) in *Pheromone Biochemistry*, eds. Prestwich, G. D. & Blumquist, G. J. (Academic, Orlando, FL), pp. 251–269.
7. Mattocks, A. R. (1972) in *Phytochemical Ecology*, ed. Harborne, J. B. (Academic, New York), pp. 179–200.
8. Rothschild, M. (1973) in *Insect/Plant Relationships*, ed. van Emden, H. F. (Blackwell, London), pp. 59–83.
9. Conner, W. E., Roach, B., Benedict, E., Meinwald, J. & Eisner, T. (1990) *J. Chem. Ecol.* **16,** 543–552.
10. Bogner, F. & Eisner, T. (1991) *J. Chem. Ecol.* **17,** 2063–2075.
11. Eisner, T. & Eisner, M. (1991) *Psyche* **98,** 111–118.
12. Conner, W. E., Eisner, T., Vander Meer, R. K., Guerrero, A., Ghiringelli, D. & Meinwald, J. (1980) *Behav. Ecol. Sociobiol.* **7,** 55–63.
13. Jain, S. C., Dussourd, D. E., Conner, W. E., Eisner, T., Guerrero, A. & Meinwald, J. (1983) *J. Org. Chem.* **48,** 2226–2270.
14. Schal, C. & Cardé, R. T. (1985) *Experientia* **41,** 1617–1619.
15. Conner, W. E., Webster, R. P. & Itagaki, H. (1985) *J. Insect Physiol.* **31,** 815–820.
16. Conner, W. E. (1985) in *Perspectives in Ethology*, eds. Klopfer, P. H. & Bateson, P. P. G. (Plenum, New York), Vol. 6, pp. 287–301.
17. Baker, T. C., Willis, M. A., Haynes, K. F. & Phelan, P. H. (1985) *Physiol. Entomol.* **10,** 257–265.
18. Dusenberry, D. B. (1989) *J. Chem. Ecol.* **15,** 971–977.
19. Conner, W. E., Eisner, T., Vander Meer, R. K., Guerrero, A. & Meinwald, J. (1981) *Behav. Ecol. Sociobiol.* **9,** 227–235.
20. Culvenor, C. C. J. & Edgar, J. A. (1972) *Experientia* **28,** 627–628.
21. Grant, A. J., O'Connell, R. J. & Eisner, T. (1989) *J. Insect Behav.* **2,** 371–385.
22. Eisner, T. (1980) in *Insect Biology and the Future*, eds. Locke, M. & Smith, D. S. (Academic, New York), pp. 847–878.
23. Bogner, F. & Eisner, T. (1992) *Experientia* **48,** 97–102.
24. Dussourd, D. E., Harvis, C. A., Meinwald, J. & Eisner, T. (1991) *Proc. Natl. Acad. Sci. USA* **88,** 9224–9227.
25. Dussourd, D. E., Ubik, K., Harvis, C., Resch, J., Meinwald, J. & Eisner, T. (1988) *Proc. Natl. Acad. Sci. USA* **85,** 5992–5996.

26. LaMunyon, C. W. & Eisner, T. (1993) *Proc. Natl. Acad. Sci. USA* **90,** 4689–4692.
27. Hare, J. F. & Eisner, T. (1993) *Oecologia* **96,** 9–18.
28. LaMunyon, C. W. (1993) Ph.D. dissertation (Cornell Univ., Ithaca, NY).
29. Sugawara, T. (1979) *J. Comp. Physiol.* **130,** 191–199.
30. LaMunyon, C. W. & Eisner, T. (1994) *Proc. Natl. Acad. Sci. USA* **91,** 7081–7084.
31. Schultz, S., Francke, W., Boppré, M., Eisner, T. & Meinwald, J. (1993) *Proc. Natl. Acad. Sci. USA* **90,** 6834–6838.
32. Eisner, T. (1988) *Verh. Dtsch. Zool. Ges.* **81,** 9–17.
33. McCormick, J. P. & Carrel, J. E. (1987) in *Pheromone Biochemistry*, eds. Prestwich, G. D. & Blumquist, G. J. (Academic, Orlando, FL), pp. 307–350.
34. Dussourd, D. E., Harvis, C. A., Meinwald, J. & Eisner, T. (1989) *Experientia* **45,** 896–898.
35. Ackery, P. R. & Vane-Wright, R. I. (1984) *Milkweed Butterflies: Their Cladistics and Biology* (Cornell Univ. Press, Ithaca, NY).
36. Brown, K. S. (1984) *Nature (London)* **309,** 707–709.
37. Boppré, M. (1986) *Naturwissenschaften* **73,** 17–26.
38. Schneider, D. (1992) *Naturwissenschaften* **79,** 241–250.
39. Arms, K., Feeney, P. & Lederhouse, R. (1974) *Science* **185,** 372–374.
40. Pivnick, K. A. & McNeil, J. N. (1987) *Physiol. Entomol.* **12,** 461–472.
41. Smedley, S. R. (1993) Ph.D. dissertation (Cornell Univ., Ithaca, NY).
42. Meinwald, J. & Eisner, T. (1995) *Proc. Natl. Acad. Sci. USA* **92,** 14–18.

The Chemistry of Signal Transduction

JON CLARDY

Extracellular molecules can influence intracellular processes, and whether we refer to this as signal transduction or chemical ecology depends on context. For example, the conjugation of two cells of the bacterial species *Streptomyces faecalis* induced by an "aggregation substance" on the surface of one and a "binding substance" on the surface of the other is a textbook example of microbial chemical ecology (1). The activation of a resting helper T cell through the stimulation of a receptor on its surface by an antigen on the surface of another cell has become one of the best studied examples of signal transduction (2, 3). In both cases a recognition between complementary elements on the surfaces of two different cells generates the potential for altered cellular function. Many other examples attest to the fundamental similarity between chemical ecology and signal transduction. (*i*) When an external molecule stimulates the IgE receptor on the surface of a mast cell, an intracellular signal leads to the release of histamine in a process referred to as degranulation (2). (*ii*) When a male gamete cell of the water mold *Allomyces* detects the sesquiterpene diol sirenin, it responds by swimming along the sirenin concentration gradient to find the female source (1). Mast-cell degranulation is usually studied as an example of signal transduction and *Allomyces* navigation as an example of chemical ecology, but in both

Jon Clardy is Horace White Professor of Chemistry at Cornell University, Ithaca, New York.

cases an external molecule dramatically affects cellular function. Our goal in studying any of these processes is the same: a molecule-by-molecule accounting of information transfer in biological systems.

While the prospect of analyzing every possible biological signal is daunting, the recognition that nature tends to use the same basic mechanism in a variety of guises makes the task less formidable. The repeated use of a common pathway is nicely illustrated by the immunosuppressive agents FK506, rapamycin, and cyclosporin. Their ability to disrupt signaling in T cells leads to immunosuppressive activity, but the same molecules that disrupt signals in T cells also prevent the degranulation of mast cells and inhibit the proliferation of yeast (2, 4). We could equally well call them antifungal, insecticidal, antiinflammatory, antiallergic, or antiretroviral as well as immunosuppressive agents (2, 4).

Another connection between cellular signal transduction and chemical ecology is the essential role played by natural products—secondary metabolites with no known role in the internal economy of the producing organism (5). Cyclosporin A (CsA), FK506, and rapamycin are all microbial natural products that are probably synthesized to chemically defend their producing organism (5). Studying these natural products as microbial chemical warfare agents would unarguably qualify as chemical ecology. However, the similarity of signaling pathways allows us to use these same natural products as probes of cellular signaling or as important chemotherapeutic agents in human disease.

The rest of this paper will focus on the factors affecting one step in one signaling pathway in resting helper T cells. The inquiry may appear overly specialized, but the strategy nature employs, using a small natural product to link two much larger proteins, has only recently been appreciated. Now that we recognize the strategy, we can expect to see it again.

SIGNAL TRANSDUCTION IN T CELLS AND THE ROLE OF NATURAL PRODUCTS

The signal to activate a resting helper T cell can be divided into three parts: the extracellular recognition of an antigen by the membrane-spanning T-cell receptor (TCR), the cytoplasmic signal transduction cascade that transmits the recognition information to the nucleus, and the activation of genes in the nucleus (3, 6). The TCR recognizes the foreign antigen, a processed peptide held in the cleft of the major histocompatibility complex (MHC) protein on the surface of the antigen-presenting cell (7, 8). Some additional interactions between T-cell surface proteins such as CD4 or CD8 with the MHC protein are needed to start the signal on its way to the nucleus. The progression of the signal from the interior portion of the TCR to the nucleus is called the cytoplasmic signal transduction cascade—a series of steps that are imperfectly understood (6). Ultimately, the signal results in the expression of a gene and the production of a gene product. The interleukin 2 (IL-2) gene is activated when the nuclear factor of activated T cells (NF-AT) binds to the IL-2 promoter region (9, 10).

Some parts of the signal process can be studied more easily than others. Much has been learned about the membrane-associated events involving MHC and TCR as well as events in the nucleus involving NF-AT and IL-2, but the cytoplasmic series of steps is less well defined. The study of cytoplasmic signal transduction, as well as a number of other biological processes, has progressed only to the extent that highly specific cell-permeant agents are available to inhibit or otherwise modify normal biological processes (11, 12). By far the most fruitful source of such agents is the realm of natural products. Natural products represent a library of tremendous chemical diversity and proven biological utility for—at the risk of tautology—only those natural products that convey some survival benefit are likely to have survived (5). Our understanding of the inhibition of the cytoplasmic signal transduction cascade of T cells originated in the discovery of three natural products: CsA, FK506, and rapamycin (Figure 1).

The discovery and utilization of CsA by Sandoz initiated a series of important advances in understanding and controlling immunosuppression (4, 13). Since 1983 this hydrophobic peptide has been widely used in clinical transplantation, and its introduction led to a remarkable increase in survival rates of transplanted livers and hearts (4). CsA (Figure 1) is produced by a variety of fungi imperfecti, notably *Tolypocladium inflatum* (formerly *Trichoderma polysporum*) isolated in Norway (13). FK506 (Figure 1) was discovered in a directed screening program at the Fujisawa Pharmaceutical Company in 1987 (14). It is produced by *Streptomyces tsukubaensis*, a species discovered in a soil sample from Tsukuba, Japan (15). While its biological effects are essentially identical to those of CsA, the two structures are chemically quite different (Figure 1). Rapamycin (Figure 1) was first described in 1975 as an antifungal agent (15–17). It was isolated from *Streptomyces hygroscopicus* from Easter Island, and the name rapamycin comes from Rapa Nui, the native name for Easter Island. All three agents have a variety of biological effects, but early studies showed that they interfere with a cytosolic signaling step and thus could be used to probe the cytoplasmic signal transduction pathway.

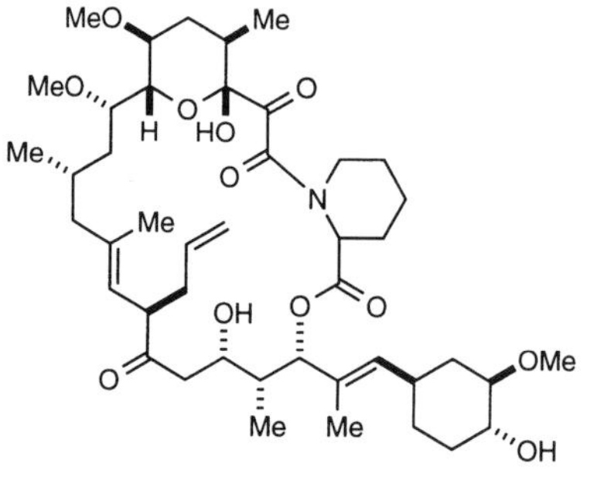

FK506

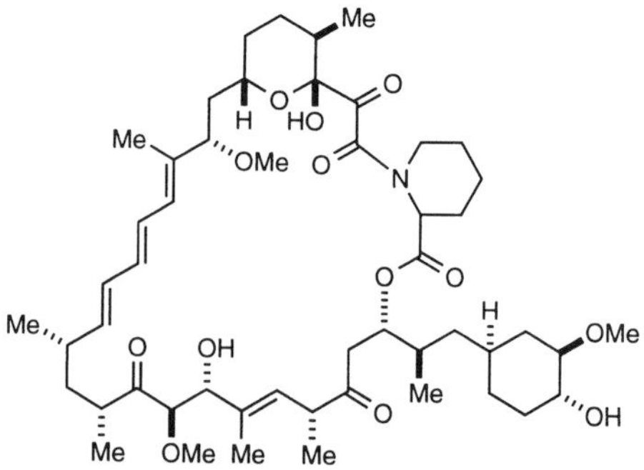

Rapamycin

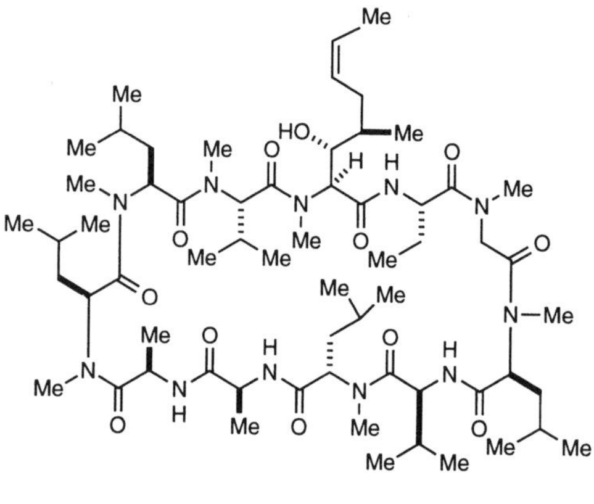

Cyclosporin A [CsA]

FIGURE 1 Structures of CsA, FK506, and rapamycin.

IMMUNOPHILINS AND THEIR COMPLEXES

The inhibitory natural products were used to identify their cellular targets, and in each case a highly specific binding protein was identified.

• A cytosolic binding protein for CsA was first isolated in 1984 and named cyclophilin, later cyclophilin A (CyPA), in reference to its high affinity for CsA (18). CyPA is a basic, abundant protein with a mass of 18 kDa, and it is found in a variety of tissues. The first clue to its function came in 1989 when two independent groups isolated the enzyme that catalyzes peptidyl proline isomerization/peptidylprolyl *cis–trans* isomerase (EC 5.2.1.8; PPIase), in protein chains and (re)discovered CyPA (19, 20). CyPA is a potent PPIase, and its enzymatic activity is strongly inhibited by CsA.

• The cellular target of FK506 was identified by two independent groups in 1989 and named FKBP, later FKBP12, for the FK506-binding protein with a mass of 12 kDa (21, 22). FKBP12 is also a potent PPIase, and FK506 strongly inhibits its activity (K_i = 0.4 nM).

• Rapamycin also appears to have a major cytoplasmic target, FKBP12 (21–23). Rapamycin binds to FKBP12 and inhibits its PPIase activity slightly better than does FK506 (K_i = 0.2 nM). FK506 and rapamycin compete for the same binding site in FKBP12.

Immunophilin is the generic term for a binding protein for an immunosuppressive agent, and all currently known immunophilins belong to the cyclophilin or FKBP families. The two families of immunophilins—like the small molecules they bind—don't have any readily apparent relationship. FKBP12 and CyPA have no sequence similarity, CsA does not inhibit FKBP12, and FK506 does not inhibit CyPA.

Since both known targets of the immunosuppressive drugs are PPIases and all known PPIases are immunophilins (24), the hypothesis that PPIase activity is involved in signal transduction and that inhibition of PPIase activity is the convergent step in immunosuppression seemed plausible. However, other studies show that the PPIase hypothesis is not tenable (6). The main lines of contradictory evidence are that (*i*) relatively low levels of drug are required for immunosuppression—drug levels at which only a fraction of the PPIase activity was eliminated; (*ii*) synthetic analogs of both FK506 and CsA are potent PPIase inhibitors but have no immunosuppressive activity; (*iii*) while both FK506 and rapamycin bind the same protein, FKBP12, they affect different stages of the T-cell cycle—only FK506 inhibits the activation step leading to IL-2 production, and (*iv*) the yeast *Saccharomyces cerevisiae*, normally sensitive to FK506 and CsA, remains viable and insensitive to either drug when FKBP or CyPA is knocked out chemically or genetically (25, 26).

In the currently accepted model, the complex of the immunosuppressive agent with its cognate protein inhibits cytoplasmic signal transduction (6). FK506, CsA, and rapamycin form complexes with immunophilins, and these complexes possess immunosuppressive activity through their ability to interact with another target. The FKBP12–FK506 and the CyPA–CsA complexes have the same target, the abundant serine-threonine phosphatase calcineurin, and thus have essentially identical biological effects (27, 28). Calcineurin is inhibited by FKBP12–FK506 and CyPA–CsA at nanomolar concentrations, but the individual components show no inhibitory activity. Thus the natural product adds a new function to its binding protein, and how this acquired function works in atomic detail has attracted several groups of investigators. FKBP12–rapamycin interacts with a protein called FRAP, which is not as well characterized as calcineurin (29).

STRUCTURAL STUDIES ON IMMUNOPHILIN–IMMUNOSUPPRESSANT COMPLEXES

Both NMR and x-ray studies have been done on free FKBP12, CyPA, and a variety of complexes (30). These studies have been reviewed recently, and the remainder of this paper will focus on relatively recent work in the FKBP area (30). Unfortunately, there are no structural studies on the FKBP12–FK506–calcineurin or CyPA–CsA–calcineurin complexes, so our understanding of the interactions is indirect and incomplete. Nevertheless, the outline, if not the complete details, of an answer is apparent.

Both NMR and x-ray structural studies on FKBP12 and its complexes show that the protein folds as a five-stranded β-sheet wrapped around a short α-helix with an overall shape resembling an ice cream cone (Figure 2 *Left*) (31–35). The ligands bind in a hydrophobic cavity between the helix and sheet in a pocket flanked by three loop regions. The protein core is composed exclusively of hydrophobic residues, many of which are highly conserved among FKBPs from different organisms. Although the protein is relatively small, it contains many structural motifs: β-sheet, α-helix, 3_{10}-helix, and an assortment of turns. The region where FK506 and FKBP12 form a composite binding surface for calcineurin is of greatest interest. The composite binding surface is the surface that interacts with calcineurin, and it must contain elements of both FK506 and FKBP12, since changes in either one can lead to complexes that do not bind to calcineurin. The loop regions near the binding pocket are of special interest: the 40s loop formed by a bulge in the fifth β-strand and the 80s loop between the second and third β-strands (36).

The high-resolution x-ray structure of the FKBP12–FK506 complex shows half of FK506's solvent-accessible surface area buried in FKBP12; the other half of the ligand is exposed. Thus, FK506 has two domains: the binding domain that interacts with FKBP12 and the effector domain that can interact with a second protein. Formation of the complex appears to have structural consequences for FKBP12. Solution NMR studies of free FKBP12 have been reported, and these studies contain valuable insights into the protein's dynamic behavior (refs. 31, 32, 37–40; S. W. Michnick, M. K. Rosen, M. Karplus and S. L. Schreiber, unpublished data). Free FKBP12 has restricted motion for the hydrophobic core, whereas several residues in the 40s and 80s loops display higher mobility. When FK506 binds to FKBP12, both the 40s and 80s loops have well-defined structures, and a picture has emerged in which the 40s and 80s loops are relatively flexible in free FKBP12 but take on a structure when FK506 is bound (36). Several studies have shown that the structure of FKBP12 shows few if any changes that depend on the bound ligand (41). Thus, the x-ray structure of the FKBP12–rapamycin complex shows no significant

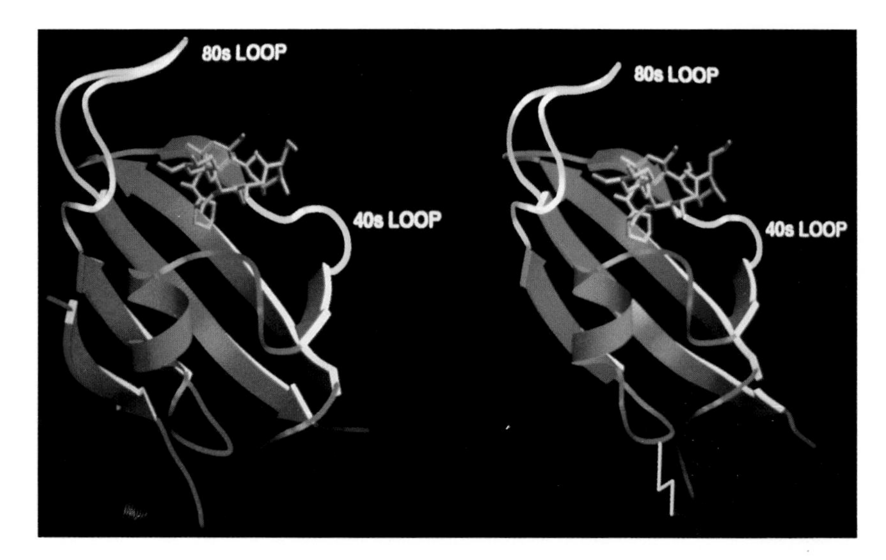

FIGURE 2 (*Left*) Diagram of the FKBP12–FK506 complex. The protein is shown by the ribbon convention and FK506 is shown as a ball-and-stick model. The β-strand numbering referred to in the text is from bottom to top. (*Right*) Diagram of the FKBP13–FK506 complex. The disulfide bridge is shown as a yellow zigzag line. The N terminus of the protein is disordered and is not shown.

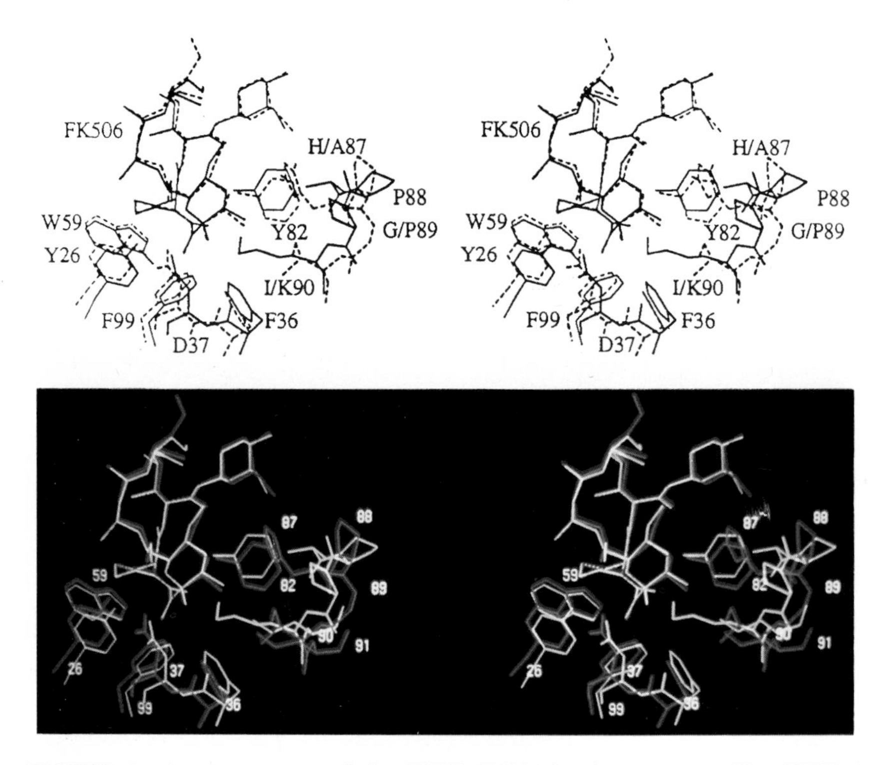

FIGURE 3 A stereo view of the FKBP–FK506 binding region. The FKBP13 structure is shown with solid (yellow) lines, and the FKBP12 structure, with dashed (blue) lines. FKBP12 numbering is used.

protein conformational changes—a conclusion reinforced by a subsequent studies (42, 43).

A detailed analysis of FKBP13, another member of the FKBP family, has led to important insights into the composite binding surface (44). FKBP13 was described by Schreiber's group in 1991 and used as the basis for an intriguing set of mutational studies (45–47). FKBP12 and FKBP13 are remarkably similar (43% amino acid identity), and the 92-amino acid C-terminal sequence of FKBP13 has 46 identical and 20 related residues when compared with FKBP12 (45). The two proteins show exact identity for all amino acids lining the FK506-binding pocket. However, embedded in this overall similarity are differences that result in a composite binding surface for FKBP13–FK506 that interacts only weakly with calcineurin (K_i = 1500 nM vs. 7.9 nM for FKBP12–FK506) (45). A series of chimeric FKBPs identified the 80s loop as the most important protein region for interaction with calcineurin. A chimeric protein with the 80s loop of FKBP13 replacing the corresponding residues of FKBP12 weakly inhibits calcineurin (K_i = 580 nM) (47). The complementary chimeric protein, in which the 80s loop of FKBP12 replaces the corresponding residues of FKBP13, is an effective calcineurin inhibitor (K_i = 27 nM) (47). Additional mutational studies patterned on the FKBP13 sequence point to the tip of the 80s loop as the key structural feature (46). The change of only two residues, proline rather than glycine at position 89 and lysine rather than isoleucine at position 90, in FKBP13 confers calcineurin-binding activity (K_i = 13 nM) (46).

Mutational studies on the 40s loop gave different, but no less interesting, results. The complete substitution of the 40s loop of FKBP12 by that of FKBP13 does not abolish activity; the resulting chimeric FKBP inhibits calcineurin (K_i = 19 nM) (47). In this chimeric protein, residue 42 is glutamine. However, the single amino acid change Arg-42 to glutamine in FKBP12 diminishes calcineurin inhibition by 2 orders of magnitude. A recently completed x-ray analysis of the FKBP13–FK506 complex adds a detailed structural understanding to these mutational results (44).

Figure 2 *Right* shows the overall structure of the FKBP13–FK506 complex, and a more detailed view comparing FKBP12–FK506 with FKBP13–FK506 is given in Figure 3. In FKBP12, Ile-90 contributes to a hydrophobic groove created by Phe-36 and the hemiketal ring of FK506 (Figure 3). In the structure of the FKBP13–FK506 complex, a lysine projects out from the 80s loop, covers this groove, and disrupts the hydrophobic pocket.

The structural analysis of the 40s loop is more complex. There are only minor differences at residues 38 and 43–45, but from residue 39 to residue 42 the FKBP13 backbone is displaced by up to 2 Å from that of FKBP12. How can the extensive modifications of the chimeric FKBP result in a calcineurin-inhibiting complex, while the single mutation at Arg-42 results in a noninhibitory complex? The available data simply don't define a unique model (12). In one model Arg-42 makes a crucial contact with calcineurin. When this contact is lost, binding is diminished. A second model assumes that Arg-42 does *not* make a contact with calcineurin; it forms part of an Arg-Asp-Tyr triad that keeps the 40s loop from interfering with calcineurin binding. The single Arg-42 mutation allows the 40s loop to interfere with binding whereas the wholesale change of residues provides a different organizational motif. The different organizational motif could include bound waters, conformational restriction from Pro-41, and an interaction between Leu-40 and His-25.

While not all the details of the composite binding surface of the FKBP12–FK506 complex are known, the general way in which FK506 adapts FKBP12 to form such a surface is becoming clear.

FK506 AND PEPTIDOMIMICRY

The discovery of FKBP12 as an abundant cytosolic protein found in a wide variety of cells suggests that it has an important role. What is that role and what is the natural ligand for FKBP12? We are exploring an approach to this teleological question based on the assumption that FK506 mimics an endogenous, probably peptidyl, substance (48).

The possibility that FK506 and rapamycin (Figure 1) mimic peptide ligands for or substrates of FKBP12 is suggested by the structural similarity between the pyranose ring, α-keto amide, and the homoprolyl fragments of FK506 on one hand and the optimal Leu-Pro and Val-Pro substrates for PPIase activity on the other (12). The peptide analogy is amplified by a pair of hydrogen bonds between FKBP12 and FK506 (Figure 4*A*) revealed in the high-resolution x-ray analysis (33, 41). This peptidomimetic analysis of FK506 leaves a stereochemical puzzle, the stereochemistry of the substituent at C-26, which has the nonnatural configuration compared with the natural peptide (note R in Figure 4*A* and R′ in Figure 4*B*).

We recently completed a high-resolution x-ray diffraction analysis of an FK506–peptide hybrid bound to FKBP12 (48). The hybrid ligands were developed around the α-keto homoprolyl moiety found in FK506—an element that appears crucial for tight binding. The adjoining amino acids were optimized to give maximum binding, and finally the variable-length tethers were added (Figure 4*C*). A 210-nm binder was selected for careful structural analysis. The resulting structure shows that many features first identified in FKBP12–FK506 persist in this hybrid structure (48). The homoprolyl ring is the most deeply buried and the tether is on the outside of the protein. The structure also has two hydrogen bonds between the

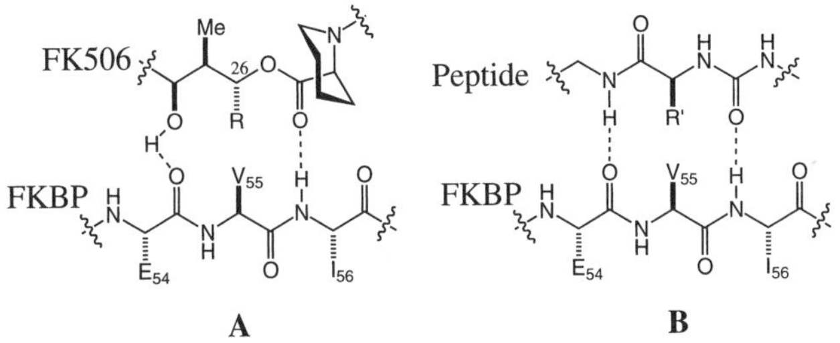

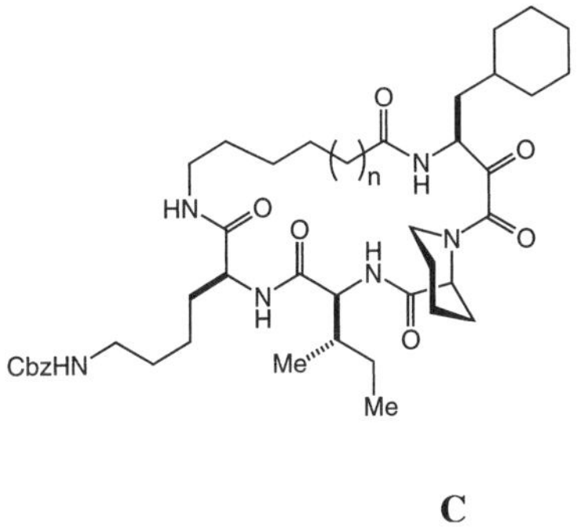

n	K_i(FKBP12)
1	1,300 nM
2	210 nM
3	13,000 nM

FIGURE 4 (*A*) Schematic of the original peptidomimetic analysis of FK506. Note the hydrogen bonds to Ile-56 and Glu-54. (*B*) Schematic of a putative peptidyl ligand binding to FKBP12, based on *A*. Note the difference in stereochemistry for R and R′. (*C*) Structure of the cyclic peptide–FK506 hybrids and their inhibitory constants for FKBP12 PPIase inhibition.

homoprolyl C═O and Ile-56 and the main-chain C═O of Glu-54 and the lysine NH of the hybrid ligand. The dipeptide fragment of the hybrid ligand binds to FKBP12 by forming a short, two-stranded antiparallel sheet. If the structures of FKBP12–FK506 and FKBP12–hybrid are superimposed, a previously unappreciated feature becomes clear. The trisubstituted double bond of FK506, which was earlier considered to be the "side-chain" substituent (R in Figure 4*A*), has an orientation and shape similar to the isoleucine amide of the hybrid. The atoms of the macrocyclic ring of FK506, which were earlier considered to be "main chain," closely follow the path of the isoleucine side chain of the hybrid. The structure suggests that the trisubstituted double bond of FK506 is an amide surrogate and its ring atoms mimic an amino acid side chain (Figure 5). Reversing the roles of these two groups in FK506 resolves the stereochemical puzzle, since the stereochemistry at C-26 now corresponds to the natural stereochemistry of an amino acid.

The structure affords other insights into the possible binding of

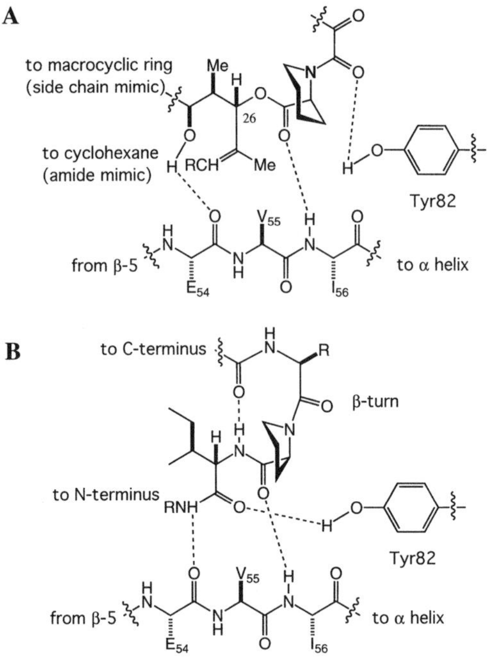

FIGURE 5 (*A*) Schematic of the revised peptidomimetic analysis of FK506. (*B*) Model of a peptide bound to FKBP12.

peptides to FKBP12. The phenolic hydrogen of Tyr-82 forms a hydrogen bond with the amide carbonyl of the dicarbonyl unit in FK506 (Figure 5*A*). In the hybrid structure, the phenolic hydrogen forms a hydrogen bond with the C═O of the isoleucine fragment (Figure 5*B*), an arrangement that would not have been possible for FK506. An intramolecular hydrogen bond from the linker carbonyl and the isoleucine NH suggests that peptides bind to FKBP12 with a *β*-turn (Fig. 5*B*), a conclusion independently reached by a theoretical analysis of the PPIase mechanism (49).

BROADER PERSPECTIVE

The desire to understand the molecular basis of information transfer in biological systems unifies many seemingly disparate disciplines, and strategies discovered in one discipline are likely to be relevant to all. The use of the natural products CsA, FK506 and rapamycin has greatly enhanced understanding of cytoplasmic signal transduction in T cells, but more importantly they have sketched an initial picture of how a small molecule can simultaneously interact with two large molecules. Chemical ecology has benefited from this conceptual framework, as demonstrated by the recent suggestion that pheromones mediate the interaction of a pheromone-binding protein with another protein target (50). We can expect additional examples.

SUMMARY

Several disciplines, including chemical ecology, seek to understand the molecular basis of information transfer in biological systems, and general molecular strategies are beginning to emerge. Often these strategies are discovered by a careful analysis of natural products and their biological effects. Cyclosporin A, FK506, and rapamycin are produced by soil microorganisms and are being used or considered as clinical immunosuppressive agents. They interrupt the cytoplasmic portion of T-cell signaling by forming a complex with a binding protein—FKBP12 in the case of FK506 and rapamycin and cyclophilin A (CyPA) in the case of cyclosporin A (CsA). This complex in turn inhibits a protein target, and the best understood target is calcineurin, which is inhibited by FK506–FKBP12 and CyPA–CsA. Mutational and structural studies help define how FK506–FKBP12 interacts with calcineurin, and the results of these studies are summarized. The existence of strong FK506–FKBP12 binding suggests that FK506 is mimicking some natural ligand for FKBP12. Synthetic and structural studies to probe this mimicry are also described.

I am grateful to Stuart L. Schreiber for generously sharing unpublished information from his laboratory. L. Wayne Schultz helped prepare the illustrations. I am grateful to the National Institutes of Health (Grant R01-CA59021) for financial support.

REFERENCES

1. Agosta, W. C. (1992) *Chemical Communication: The Language of Pheromones* (Freeman, New York), pp. 14–18, 29–31.
2. Rosen, M. K. & Schreiber, S. L. (1992) *Angew. Chem. Int. Ed. Engl.* **31,** 384–400.
3. Sigal, N. H. & Dumont, F. J. (1992) *Annu. Rev. Immunol.* **10,** 519–560.
4. Fliri, H., Bauman, G., Enz, A., Kallen, J., Luyten, M., Mikol, V., Movva, R., Quesniauz, V., Schreier, M., Walkinshaw, M., Wenger, R., Zenke, G. & Zurini, M. (1993) *Ann. N.Y. Acad. Sci.* **696,** 39–47.
5. Williams, D. H., Stone, M. J., Hauck, P. R. & Rahman, S. K. (1989) *J. Nat. Prod.* **52,** 1189–1208.
6. Schreiber, S. L. (1991) *Science* **251,** 283–287.
7. Stern, L. J., Brown, J. H., Jardetzky, T. S., Gorga, J. C., Urban, R. G., Strominger, J. L. & Wiley, D. C. (1994) *Nature (London)* **368,** 215–221.
8. Chicz, R. M. & Urban, R. G. (1994) *Immunol. Today* **15,** 155–159.
9. Emmel, E. A., Verweij, C. L., Durand, D. B., Higgins, K. M., Lacy, E. & Crabtree, G. R. (1989) *Science* **246,** 1617–1620.

10. Flanagan, W. F., Corthesy, B., Bram, R. J. & Crabtree, G. R. (1991) *Nature (London)* **352,** 803–807.
11. Schreiber, S. L., Albers, M. W. & Brown, E. J. (1993) *Acc. Chem. Res.* **26,** 412–420.
12. Rosen, M. K. (1993) Ph.D. dissertation (Harvard Univ., Cambridge, MA).
13. Rüegger, A., Kuhn, M., Lichti, H., Loosli, H.-R., Huguenin, R., Quiquerez, C. & von Wartburg, A. (1976) *Helv. Chim. Acta* **59,** 1075–1092.
14. Tanaka, H., Kuroda, A., Marusawa, H., Hatanaka, H., Kino, T., Hoto, T. & Hashimoto, M. (1987) *J. Am. Chem. Soc.* **109,** 5031–5033.
15. Vezina, C., Kudelski, A. & Sehgal, S. N. (1975) *J. Antibiot.* **28,** 721–726.
16. Sehgal, S. N., Baker, H. & Vezina, C. (1975) *J. Antibiot.* **28,** 727–732.
17. Swindells, D. C. N., White, P. S. & Findlay, J. A. (1978) *Can. J. Chem.* **56,** 2491–2492.
18. Handschumacher, R. E., Harding, M. W., Rice, J., Drugge, R. J. & Speicher, D. W. (1984) *Science* **226,** 544–547.
19. Fischer, G., Wittman, L. B., Lang, K., Kiefhaber, T. & Schmid, F. X. (1989) *Nature (London)* **337,** 476.
20. Takahashi, N. T., Hayano, T. & Suzuki, M. (1989) *Nature (London)* **337,** 473–475.
21. Harding, M. W., Galat, A., Uehling, D. E. & Schreiber, S. L. (1989) *Nature (London)* **341,** 758–760.
22. Siekierka, J. J., Hung, S. H. Y., Poe, M., Lin, C. S. & Sigal, N. H. (1989) *Nature (London)* **341,** 755–757.
23. Fretz, H., Albers, M. W., Galat, A., Standaert, R. F. & Schreiber, S. L. (1991) *J. Am. Chem. Soc.* **113,** 1409–1411.
24. Rahfeld, J.-U., Schierhorn, A., Mann, K. & Fischer, G. (1994) *FEBS Lett.* **343,** 65–69.
25. Wiederrecht, G., Brizuela, L., Elliston, D., Sigal, N. & Siekierka, J. (1991) *Proc. Natl. Acad. Sci. USA* **88,** 1029–1033.
26. Tropschug, M., Barthelmess, I. & Neupert, W. (1989) *Nature (London)* **342,** 953–955.
27. Liu, J., Farmer, J. D., Lane, W. S., Friedman, J., Weissman, I. & Schreiber, S. L. (1991) *Cell* **66,** 807–815.
28. Liu, J., Albers, M., Wandless, T. J., Luan, S., Alberg, D. A., Belshaw, P. J., Cohen, P., MacKintosh, C., Klee, C. B. & Schreiber, S. L. (1992) *Biochemistry* **31,** 3896–3901.
29. Brown, E. J., Albers, M. W., Shin, T. B., Ichickawa, K., Keith, C. T., Lane, W. S. & Schreiber, S. L. (1994) *Nature (London)* **369,** 756–758.
30. Clardy, J. (1994) *Perspect. Drug Discovery Des.* **2,** in press.
31. Moore, J. A., Peattie, D. A., Fitzgibbon, M. J. & Thomson, J. A. (1991) *Nature (London)* **351,** 248–250.
32. Michnick, S. W., Rosen, M. K., Wandless, T. J., Karplus, M. & Schreiber, S. L. (1991) *Science* **251,** 836–839.
33. VanDuyne, G. D., Standaert, R. F., Karplus, P. A., Schreiber, S. L. & Clardy, J. (1991) *Science* **251,** 839–842.
34. VanDuyne, G. D., Standaert, R. F., Schreiber, S. L. & Clardy, J. (1991) *J. Am. Chem. Soc.* **113,** 7433–7434.
35. Weber, C., Wider, G., Freyberg, B. v., Traber, R., Braun, W., Widmer, H. & Wüthrich, K. (1991) *Biochemistry* **30,** 6563–6574.
36. Clardy, J. (1993) *Ann. N.Y. Acad. Sci.* **685,** 37–46.
37. Lepre, C. A., Thomson, J. A. & Moore, J. M. (1992) *FEBS Lett.* **302,** 89–96.
38. Meadows, R. P., Nettesheim, D. G., Xu, R. T., Olejniczak, E. T., Petros, A. M.,

Holzman, T. F., Severin, J., Gubbins, E., Smith, H. & Fesik, S. W. (1993) *Biochemistry* **32,** 754–765.

39. Petros, A. M., Gampe, R. T., Jr., Gemmecker, G., Neri, P., Holzman, T. F., Edalji, R., Hochlowski, J., Jackson, M., McAlpine, J., Luly, J. R., Pilot-Matiaas, T., Pratt, S. & Fesik, S. W. (1991) *J. Med. Chem.* **34,** 2925–2928.
40. Petros, A. M., Luly, J. R., Liang, H. & Fesik, S. W. (1993) *J. Am. Chem. Soc.* **115,** 9920–9924.
41. VanDuyne, G. D., Standaert, R. F., Karplus, P. A., Schreiber, S. L. & Clardy, J. (1993) *J. Mol. Biol.* **229,** 105–124.
42. Becker, J. W., Rotonda, J., McKeever, B. M., Chan, H. K., Marcy, A. I., Wiederrecht, G., Hermes, J. D. & Springer, J. P. (1993) *J. Biol. Chem.* **268,** 11335–11339.
43. Holt, D. A., Luengo, J. I., Yamashita, D. S., Oh, H.-J., Konialian, A. L., Yen, H.-K., Rozamus, L. W., Brandt, M., Bossard, M. J., Levy, M. A., Eggleston, D. S., Liang, J., Schultz, L. W., Stout, T. J. & Clardy, J. (1993) *J. Am. Chem. Soc.* **115,** 9925–9938.
44. Schultz, L. W., Martin, P. K., Liang, J., Schreiber, S. L. & Clardy, J. (1994) *J. Am. Chem. Soc.* **116,** 3129–3130.
45. Jin, Y.-J., Albers, M. W., Lane, W. S., Bierer, B. E., Schreiber, S. L. & Burakoff, S. J. (1991) *Proc. Natl. Acad. Sci. USA* **88,** 6677–6681.
46. Rosen, M. K., Yang, D., Martin, P. K. & Schreiber, S. L. (1993) *J. Am. Chem. Soc.* **115,** 821–822.
47. Yang, D., Rosen, M. K. & Schreiber, S. L. (1993) *J. Am. Chem. Soc.* **115,** 819–820.
48. Ikeda, Y., Schultz, L. W., Clardy, J. & Schreiber, S. L. (1994) *J. Am. Chem. Soc.* **116,** 4143–4144.
49. Fischer, S., Michnick, S. W. & Karplus, M. (1993) *Biochemistry* **32,** 13830–13837.
50. Du, G., Ng, C.-F. & Prestwich, G. D. (1994) *Biochemistry* **33,** 4812–4819.

Chemical Signals in the Marine Environment: Dispersal, Detection, and Temporal Signal Analysis

JELLE ATEMA

Chemical signals identify biologically important targets for those who have the proper receivers. We assume that selection pressure can act on both the biochemical and the physiological regulation of the signal and on the morphological and neurophysiological filter properties of the receiver. Communication is implied when signal and receiver evolve toward more and more specific matching, culminating in well-known sex pheromone systems. In other cases, receivers respond to portions of a body odor bouquet that is released to the environment not as a (intentional) signal but as an unavoidable consequence of metabolic activity or tissue damage. Breath, sweat, urine, feces, their aquatic equivalents, and their bacterial and other symbiotic embellishments all can serve as identifiers for chemoreceptive animals interested in finding food or hosts. Body fluids released from damaged tissues and decay products from dead organisms can be particularly potent signals. Since all organisms must release metabolites in order to live, and since any such release is a potential target of opportunity for predators and parasites, one may expect that several forms of chemical camouflage have evolved to obscure one's chemical presence. Both communication signals and camouflage depend on signal-to-background contrast or lack thereof: communication emphasizes contrast; camouflage works toward lower contrast.

Jelle Atema is professor of biology and director of the Boston University Marine Program at the Marine Biological Laboratory, Woods Hole, Massachusetts.

Contrast can be provided by chemical specificity of the signal (spectral contrast) and by temporal changes in concentration of the signal (dynamic temporal contrast). Spectral contrast is created by unique compounds and by unique mixtures of compounds, including ordinary ones. Temporal contrast emerges from the rate at which the concentration of a compound changes with time, including the repetition rate. Temporal changes reflect spatial patchiness and hold information for chemotactic behavior at different spatiotemporal scales. The two classical methods of camouflage, well known in the visual signal world, may also operate in the chemical signal world, although they are virtually unstudied. To avoid detection, animals with visual predators hide and remain motionless, or they look and move like their background; animals with chemically hunting predators may build impermeable shells and store urine and feces until it is safe to release them, or they may produce metabolites that match the environment in mixture composition and temporal distribution.

Unlike wave or wave-like propagation of acoustic, visual, and other electromagnetic signals, chemical signals disperse through the environment by molecular diffusion and bulk flow. At small spatial scales—in practice below 10 μm—diffusion is a biologically useful transport mechanism and, given the constraints of viscous fluid boundary layers, often the only effective mechanism. At larger scales, flow is necessary to obtain metabolic energy (e.g., oxygen, food particles), to eliminate wastes (e.g., carbon dioxide, urine), and to send and receive chemical signals. The constraints of metabolism and sensory information are probably different, so that we could expect animals to generate separate metabolic currents and information currents. In practice, they may use the same current-generating mechanisms and then control the currents and the chemical composition of these currents to serve different functions at different times. Controlling the timing, velocity, and direction of information currents is important whether they are used to send or receive chemical signals. Animal-generated currents can be laminar at small scales ($<$1 cm) or turbulent at larger scales. Both include the possibility of temporal information. In this paper I will focus on temporal information in marine chemical signals and on the use of urine dispersal in chemical communication.

The marine environment is filled with sources of chemical signals in a wide range of overlapping spatial scales (1), from the metabolites of a single marine bacterium (diameter, $<10^{-6}$ m) to the odor plumes left behind a traveling school of tuna (school size, $>10^2$ m) or emanating from a whale carcass (plume size, $>10^3$ m). Constrained by physics, chemistry, and biology, chemical signals have a finite lifetime. When released into the environment, they disappear below detectable levels as a result of turbulent mixing, molecular diffusion, adsorption, photolysis, and chemical transformation and through uptake and breakdown by bacteria, microorganisms, small invertebrates, and invertebrate larvae. During their lifetime, chemical signals exist as patches of constantly varying sizes and shapes. Temporal scales tend to follow spatial scales: molecular diffusion would obliterate a 10-μm (diameter) spherical patch of small dissolved molecules in 10 s (i.e., concentration drop to 0.1% of original) and a 1000-m-long surface oil slick resulting from a whale carcass being devoured by sharks can be visible for days.

Since all organisms during their lifetime experience only a limited range of spatiotemporal scales, it is important to understand the scales that are relevant to particular animals if we are to discover the mechanisms animals use to extract information from the dispersal patterns of chemical signals. Signal longevity and temporal pattern are no less important in chemical signals than they are in other sensory stimuli. The speed of temporal signal detection and processing depends on the encounter rate with odor patches and their spatial gradients. This encounter rate is determined by the speed of search behavior. Perhaps as a result of this correlation we tend to see a dependence on chemical signals in "slow" animals—e.g., crustacea, mollusca (but not visually hunting squid). Needless to say, many slow animals can have very fast escape responses, thus linking their slow search to sensory processing, not to intrinsic locomotion limitations. In lobsters, chemical gradient search reduced their normal walking speed by half, and their chemoreceptive flicker fusion rate (i.e., the stimulus pulse frequency that causes responses to fuse) is an order of magnitude lower than typical visual fusion rates.

In order to detect chemical signals in the sea, receptors must recognize them against a background of many other chemical compounds. This involves not only receptor specificity and diversity but also recognition of the intensity and time course of the signal to allow a receptor to distinguish a true signal from random events. A point receptor moving across a field of patches will see them as a chaotically fluctuating intensity pattern. We assume therefore that receptors evolved with both spectral and temporal properties tuned to signal recognition and that the tuning properties of receptor organs reflect, on the one hand, the constraints imposed by prevalent natural stimulus conditions and, on the other hand, the demands for specific information most useful for that animal's behavioral tasks. In a true sense, the receptor organ determines by its spectral and temporal tuning what is signal and what is noise. The same environmental signal distribution can therefore yield different information as a result of tuning. This applies to different species as well as to different organs of a single species. Note that many species have a multitude of different chemoreceptor organs, mostly unstudied physiologically and behaviorally. By comparison, humans have three, perhaps four, chemoreceptor organs serving the senses of smell, taste, trigeminal, and perhaps vomeronasal chemoreception, each tuned to a different spectrum of stimuli and to different encounter rates.

A full understanding of chemical ecology must therefore include not only the characterization of chemically unique signals but also their environmental dispersal and degradation patterns that are an intrinsic part of chemosensory transduction and signal processing and lead to the appropriate behavioral responses. One of the most difficult tasks has been to measure the natural stimulus dispersal patterns at a spatial and temporal resolution relevant to the animal.

In this paper I will discuss a single species to illustrate some of the processes of marine chemical signal dispersal, receptor tuning, and chemotactic and social behavior. In an effort to understand the underwater world of chemical signals—so foreign to humans—we attempt to see the marine environment through the many chemosensory organs of the lobster, *Homarus americanus*, an animal that has demonstrated its ability to communicate with chemical signals, urine release, and a variety of information currents and to extract spectral and temporal chemical information from its turbulent environment (2). To investigate the spatiotemporal dynamics of natural odor dispersal that are important physiologically and behaviorally, aquatic chemical signals and large arthropods offer significant advantages. Due to the greater density of water, the relevant scales of turbulence are about an order of magnitude smaller in water than in air, facilitating behavioral experiments and odor dispersal measurements. Chemical signals in solution can be quantified and expressed in molarity. Electrochemical microelectrodes can be constructed to measure marine chemical stimulus patterns at the proper spatial scale of receptor sensilla, and recent advances in signal processing have allowed the temporal resolution to exceed the animal's (3). These electrodes mimic crustacean chemoreceptor organs, which are built with cuticular pegs that form distinct boundaries with the surrounding fluid, uncomplicated by mucus transitions. Large arthropods can easily carry instrumentation packets, including catheters, electrodes, amplifiers, and transmitters.

SIGNAL RECEPTION AND ANALYSIS: TEMPORAL ANALYSIS OF CHEMICAL SIGNALS TO DETERMINE SPATIAL GRADIENTS

This section reviews four different experimental approaches that together argue in favor of a temporal analysis function of lobster olfaction. The experiments include high-resolution measurements of turbulent odor dispersal and lobster sampling behavior, electrophysiological recording of *in situ* single cell responses to controlled and chaotic stimuli, and behavioral analysis of orientation and localization of odor sources.

Spatial Gradients in Turbulent Odor Plumes

Since turbulent odor plumes simultaneously contain a broad suite of eddy sizes, the first problem is to decide which range of frequencies to measure and at which spatiotemporal scale to resolve the signal/noise complex. For spatial resolution, we scaled our odor tracer sensor (carbon-filled glass microelectrode) to the physical dimension of the lobster's 30-μm (diameter) olfactory receptor sensillum. For temporal resolution, we chose 5 ms [IVEC-10 (*in vivo* electrochemistry analyzer), and the Gerhardt custom package, Medical Systems Corp., Greenvale, NY], exceeding both the lobster's olfactory flicker-fusion frequency of 4 Hz (see below) and laboratory plume frequencies, which have little energy above 40 Hz. We studied odor dispersal patterns resulting from biologically scaled, constantly emitting jet sources in slow background flow. From high-resolution plume measurements using dopamine as a tracer, we described the spatial distribution of encounter probabilities of eddy features such as peak concentration, concentration gradients at their leading edge, intermittency, etc. (4). Some of these stimulus features showed spatial gradients that could be used to track and locate an odor source. This demonstrates that at the measuring scale of animal sensors, purely chemoreceptive information can be extracted from which the direction of a distant odor source can be estimated. In other words, true chemotaxis based on temporal analysis of odor patch features is theoretically possible.

Orientation and Navigation in Odor Plumes

Subsequent behavioral experiments in the same laboratory flume (5) showed that lobsters indeed located the source of food odor plumes, whereas prior lesion experiments had shown that distance orientation is guided by olfaction (6) and the final approach of about 30 cm by taste (7). Smell and taste are defined functionally (8). The olfactory orientation was composed of an initial scanning phase (characterized by low walking speed and large heading angles) followed by an increasingly fast and accurate approach. Both speed and accuracy reached a maximum that was maintained until walking legs (taste) took over the final search. Curiously, when not engaged in chemotactic search, lobsters can walk twice as fast. Unlike some insects (9, 10), lobsters do not appear to use innate motor programs, such as counterturning. Instead, the individuality of their tracks and their relatively slow top speed during chemotactic approach suggest that they monitor and follow the variable spatial gradients characteristic of turbulent plumes. If walking animals such as lobsters use odor patch information for plume navigation, then free-swimming animals without ground reference will be even more likely to extract directional cues from turbulent odor dispersal patterns.

Two questions arise from this result. Do lobsters use only chemical and not mechanosensory information, and why do lobsters not use ground reference and head up-current? Since turbulent odor dispersal is based on water flow patterns, we must investigate the role of microflow patterns in plume orientation behavior. As for ground reference, we speculate that the flow patterns of the lobster's natural environment may be too complex to allow for efficient rheotactic behavior in odor source localization. This complexity is most likely caused by a mismatch between turbulent scales and animal body size and sampling scales.

Odor Sampling Behavior

Feature extraction is the primary function of sensory systems. To this end, sense organs almost always use three different levels of signal filtering: a physical filter based on receptor organ morphology and associated behavior, a receptor cell filter based on biophysical and biochemical properties of transduction and adaptation, and a neural filter based on network connectivity in the central nervous system. We investigated the first two filters for lobster olfaction. The olfactory sensilla ("hairs") known to be critical for efficient orientation behavior (6) form a dense "toothbrush" on the distal half of the lateral flagellum of the antennules. Video analysis and high-resolution electrochemical measurements showed that under low flow conditions (<5 cm/s) this brush forms a dense boundary layer which traps existing odor and shields the receptors from rapid odor access, thus making smelling virtually impossible (3). Thus, lobsters must flick their antennules (i.e., sniff) to smell. Flicking behavior drives water at high velocity (>12 cm/s) through the brush and causes the hairs to tremble in their sockets. This allows rapid odor exchange around the entire 1-mm-long shafts of all the hairs (3). Flick rates of up to 4 Hz occur in excited lobsters (11). The bilateral antennules allow for spatial comparison essential for efficient orientation (6).

Temporal Resolution of Olfactory Receptor Cells

Lobster chemoreceptor cells show a great diversity of filter properties. Electrophysiological measurements show that each receptor cell is tuned not only to one or a few preferred compounds (12, 13) but also to a preferred frequency (14, 15). Temporal resolution in chemoreception is affected by at least five different stimulus parameters: rate of stimulus concentration increase (= pulse slope), amplitude and duration of a single odor pulse, pulse repetition rate, and pulse-to-background concentration ratio. Pulse slope corresponds to the arrival of an odor-flavored eddy, duration corresponds to its size; repetition rate represents arrival of different eddies and their spacing. All five stimulus parameters depend on the degree of turbulent and diffusive mixing and thus correlate with time since release and distance from the source. The dynamic response properties of receptor cells are determined by two somewhat independent cellular processes: adaptation and disadaptation (or recovery). A cell's adaptation rate determines its preferred stimulus slope and duration, and its recovery rate determines its flicker fusion frequency. Results from qualitative experiments in which odor and single cell responses were measured simultaneously and with high spatiotemporal resolution indicate that steep pulse slope and large interpulse interval are important excitatory stimulus features. Unfortunately, the nonlinear dynamics of adaptation and disadaptation processes preclude a simple solution for determining meaningful transfer functions. Therefore, stimulus features must be analyzed one at a time to measure their effect on receptor cell responses, and—ultimately—to reconstruct the temporal analysis capabilities of the lobster olfactory organ.

Carefully controlled square pulses of odor with identical amplitude and different pulse duration (50–1000 ms) resulted in response maxima for 200-ms stimuli; longer pulse durations did not result in greater responses as measured by spike firing frequency. This reflects the effects of cellular adaptation mechanisms—presumably receptor phosphorylation—that begin to overwhelm excitatory processes at about 200 ms. For all stimulus pulse amplitudes, adaptation is complete at about 400 ms. Thus, responses to a single pulse are determined largely by initial stimulus pulse parameters and by the cell's intrinsic physiological properties and negligibly by pulse features beyond 200 ms. Flicker fusion to 100-ms square pulses occurs at 4 Hz for the fastest receptor cells (15). Some insect receptor cells can follow 4- to 10-Hz pulse rates (16–18). These results indicate that cellular disadaptation processes require at least 150 ms before the cell can respond again to phase-lock with subsequent pulses of similar magnitude. In addition, pulse amplitude-to-background ratio determines response magnitude: backgrounds shift the response function to pulses toward higher concentrations.

Discussion

The results demonstrate (*i*) that in turbulent odor plumes, purely chemical spatial gradients can be calculated when measuring with sensors scaled to lobster olfactory organs, (*ii*) that rapid odor access to the lobster's olfactory organs (under low ambient flow conditions) is accomplished by flicking, (*iii*) that the maximum observed flick rate of 4 Hz corresponds to the neurophysiologically determined flicker fusion frequency of olfactory receptor cells, (*iv*) that receptor cells are tuned not only to specific compounds but also to different temporal parameters of odor, and (*v*) that lobsters locate odor sources with search paths, walking speed, and turning behavior suggestive of chemotaxis based on patchy odor distributions. These results show that turbulently dispersed odor patches from a single source contain spatial gradients of signal parameters and that some of these parameters are recognized by olfactory temporal filters. We hypothesize that these temporal odor parameters provide information for chemotactic navigation in odor plumes. At this point, we have learned some of the physiologically determined temporal filters, but we do not know the exact nature of the behaviorally relevant signal features nor the sampling regimes and signal processing required to lead to efficient navigation. We are approaching these questions with behavioral, physiological, computational (19), and robotic methods (20). In the context of pheromone communication, Bossert (21) calculated that amplitude modulation could enhance information transmission very significantly. Some moths release pheromone in puffs with 1-s periodicity (22) and it was found that Bossert's calculations must be reevaluated in light of current knowledge of temporal filter properties of chemoreceptor cells and the unpredictability factor of turbulent dispersal that characterizes most natural environments.

SIGNAL PRODUCTION AND BROADCASTING: URINE DISPERSAL IN CHEMICAL COMMUNICATION

This section reviews the complex currents lobsters generate to eliminate metabolites and broadcast chemical signals and the return currents from which they obtain chemical signals and metabolic energy. Lobsters are examples of hard-shelled animals that store urine and feces, allowing them to be chemically "quiet" when necessary.

Information Currents and Urine Signals

H. americanus utilizes three current-generating mechanisms that can operate separately or in combination; all three are implicated in chemical communication. The scaphognatites inside the gill chambers generate a powerful gill current which jets forward from bilateral "nozzles." This current reaches distances of up to seven body lengths in adults (2) and velocities of 3 cm/s near the nozzle. It is usually a bilateral current and it carries the animal's gill metabolites. Mature-sized lobsters under summer temperatures rarely cease producing this breathing current; in winter, the current stops for episodes of several seconds, presumably reflecting the animal's lower metabolic demands. In addition, urine can be released into this current from bilateral bladders through small ventrally directed nephropores at the base of the antennae. Lobsters appear to release into the urine the products of a small cluster of glands located in the nephropore nipple some 100 μm inside the excretory pore (23). Both the gland with its duct and surrounding muscle tissue and the muscular valve of the nephropore appear designed to give the animal control over chemical signals released into the gill current. Glandular products and function are unknown, but morphology suggests that it is a rosette gland related to common tegumental glands (24), and histological stains suggest that its product is proteinaceous, carbohydrate, or glycoprotein (25).

Further control of signaling is possible through a redirecting of the gill current by the exopodites of the three maxillipeds. It appears that the exopodite of the first maxilliped can be positioned to—partially—cover the gill chamber outflow nozzle, thus deflecting and redirecting forward water flow. The large feathery exopodites of the second and third maxillipeds then fan the deflected water backwards, while drawing in a slow flow of water from around the animal's head (2). The antennules flick and thus sample odor within this area, the radius of which is about the length of the antennule. The exopodite fan current represents the second lobster-generated current. It can be bilateral or unilateral on either side.

Together, the two lobster-generated currents that can be measured around the animal's anterior end are complex and carefully controlled. They are ideally suited to carry urine, urine pheromones, and gill metabolites away from the lobster to specified directions. Simultaneously, the water displaced by these outgoing currents results in incoming currents with chemical signals from the environment that can be sampled by the antennular chemoreceptors.

The third and most powerful lobster-generated current is the pleopod current, which draws water from below the lobster and blows it posteriorly (2). Typically, the lobster raises its tail and beats its pleopods to generate this current, which is sufficiently powerful in adult animals to help the animal in forward motion and in climbing onto rocks. In smaller animals, particularly in IVth stage post-larvae, this current is used for forward swimming and food particle capture (26). Cohabiting mature males pleopod-fan frequently at the entrance of their shelter, thereby sending male odor and the odor of the cohabiting female into the environment. In the laboratory, this behavior results in greatly increased visits to his shelter by premolt females and other lobsters (27). The signals involved are unknown but could easily include urine-carried pheromones.

Urine Signals in Dominance and Courtship

Males compete for dominance, which they establish during fights. Experiments with chronic catheters show that aggressive animals release pulses of urine during the fight and that lobsters stop urine release as soon as they have lost (28, 29). Urine release was not observed when lobsters were disturbed by the experimenter, thus implying that urine release by aggressive animals is social and carefully controlled (28). During subsequent encounters, as long as a week (but less than 2 weeks) later, the opponents remember each other, as evidenced by the fact that almost no fight takes place: the former loser avoids the winner (30). Chemical signals in the urine are involved in this memory, as demonstrated by experimental evidence that both catheterization of urine release and temporary lesion of antennular chemoreceptors result in renewed fights on subsequent days (29, 30).

After dominance is established, the dominant male occupies a "preferred" large shelter which becomes a focus of social interactions. Mature, premolt females visit frequently (31, 32). For cohabitation and subsequent mating, females in naturalistic aquaria chose the dominant male over subdominants (31). In choice tests females prefer larger dominant males (23). Females make these behavioral decisions both from a distance and at the shelter entrance. Discrimination by females is lost when males are catheterized and can be regained when that male's urine is artificially released near the male (23). Thus, male urine cues for female choice are implied but have not been identified.

Visiting females stand still at the entrance of the male shelter for many seconds. Observation of their expodites shows that they alternate between fanning and not fanning. Therefore, when not fanning, they blow their gill current into the male shelter. At the time of female visits, the male often stands inside and away from the female entrance, flicking his antennules, fanning his exopodites (thus drawing water toward his antennules and redirecting his gill current backwards), and occasionally fanning his pleopods—all resulting in female odor reaching the male. In an agonistic context, field (32) and laboratory observations show that a female has to be considerably larger to force a male out of his shelter. In a courtship context, males accept even much smaller premolt females and cohabit with them, one at a time, each for 1–2 weeks. Although such context appears to be provided by chemical signals (33, 34), female sex pheromones have not been chemically identified. We speculate that special glandular products in the urine may be involved to identify the visiting female as a mature and premolt female lobster. Since both males and females at all molt stages possess active nephropore glands (23), its product would have to be modified for different sexes and state of maturity if it were to be used in the contexts described above. Alternatively, changes in perception of the signal could be involved.

Cohabitation lasts from a few days to weeks. The female molts during this time, and mating follows the female molt after one-half hour. Over the premolt cohabitation period, male pleopod fanning increases. It reaches its maximum during the days of female molting and wanes in the postmolt cohabitation days. Fanning males often stand at one of their shelter entrances with their abdomen raised high and slightly outside the entrance. This results in a strong current running through and out of his shelter (35). This current contains all male and female metabolites (including possible pheromones) released by the cohabiting pair. Male fanning is positively correlated with visits of other lobsters, including premolt females, to the shelter. These other premolt females do not molt but wait for their turn to cohabit with the dominant male (27). Female molt staggering implies control of the female molt cycle. We speculate that the male pleopod current, which serves as an advertisement device to attract lobsters ("releaser pheromones;" ref. 36), also contains molt inhibiting signals ("primer pheromones;" ref. 36) for visiting females who are released from this inhibition when they start cohabiting with the dominant male leading to molting and mating within days.

Discussion

All indications are that we are only just beginning to see a few threads of the rich fabric of chemical signals that link lobsters to each other and to their environment. Exoskeleton, bladders, glands, and control of currents all indicate that these animals can be chemically quiet and release specific signals at critical times during aggression and courtship. Chemical signals appear to be used to remember individuals and to facilitate stable dominance hierarchies.

One may wonder why lobsters appear to use urine as a dispersal solvent for chemical signals, whereas terrestrial arthropods such as the well-studied insects use direct release of gland products into the air. Perhaps the answer is that small animals in air (such as insects) are always in danger of desiccation. By contrast, marine lobsters and crabs are relatively large and may experience only minor water loss problems due to osmosis. Thus, it may not be difficult for a 500-g lobster to store 10 ml of urine and release it during a dominance battle at a rate of up to 1 ml/min (27). The advantage of urine-carried pheromones is that the dispersal mechanism already exists: urine is injected into the gill current, which in turn injects into ocean currents.

In conclusion, lobsters are excellent models to learn about the natural world of odor dynamics, but they are probably not unique: they are opening our minds to possibilities not yet explored with other animals. The odor environment is richer and more complex than we know today and lobsters are showing us several thus far unexplored dimensions: navigation based on the characteristics of rapid odor signals, information currents, urine-based pheromone broadcasting in the sea, and—possibly—chemical camouflage.

SUMMARY

Chemical signals connect most of life's processes, including interorganismal relationships. Detection of chemical signals involves not only recognition of a spectrum of unique compounds or mixtures of compounds but also their spatial and temporal distribution. Both spectral and temporal signal processing determine what is a signal and what is background noise. Each animal extracts its unique information from the chemical world and uniquely contributes to it. Lobsters have provided important information on temporal signal processing. Marine chemical signals can be measured with high spatio-temporal resolution giving us a novel view of the lobster's environment. Lobster chemoreceptor cells have flicker fusion frequencies of 4 Hz and can integrate stimuli over 200 ms, closely corresponding to odor sampling behavior with 4-Hz "sniffs." Using this information, spatial odor gradients can be determined from temporal analysis of odor patches typical of turbulent dispersal. Lobsters appear to use this information to locate odor sources. Lobster social behavior depends greatly on chemical signals. Urine carries important information for courtship, dominance, and individual recognition. A novel gland in the nephropore is strategically located to release its products into the urine. Urine, in turn, is injected into the gill current, which jets water 1–2 m ahead of the animal. Lobsters control three different currents that carry chemical signals to and from them. The study of odor dynamics has only just begun. It will be exciting to see how signal dispersal, receptor temporal tuning, neural processing, and animal behavior interact to enhance signals for communication and detection and to reduce signals for chemical camouflage.

I thank my students, postdoctoral associates, and colleagues for their many contributions to the exciting discoveries made in the chemical world of lobsters. For recent contributions I mention specifically my long-time colleague Dr. Rainer Voigt, my former students Dr. Paul Moore and Dr. George Gomez, my postdoctoral associates Drs. Thomas Breithaupt and Jennifer Basil, and my students Paul Bushmann and Christy Karavanich. This work was supported by National Science Foundation Grants IBN-9212650 and IBN-9222774 and National Institutes of Health Grant 5 PO1NS25915.

REFERENCES

1. Atema, J. (1988) in *Sensory Biology of Aquatic Animals*, eds. Atema, J., Popper, A. N., Fay, R. R. & Tavolga, W. N. (Springer, New York), pp. 29–56.
2. Atema, J. (1985) *Soc. Exp. Biol. Symp. Ser.* **39,** 387–423.
3. Moore, P. A., Atema, J. & Gerhardt, G. A. (1992) *Chem. Senses* **16,** 663–674.
4. Moore, P. A. & Atema, J. (1991) *Biol. Bull.* **181,** 408–418.
5. Moore, P. A., Scholz, N. & Atema, J. (1991) *J. Chem. Ecol.* **17,** 1293–1307.
6. Devine, D. V. & Atema, J. (1982) *Biol. Bull.* **163,** 144–153.
7. Derby, C. D. & Atema, J. (1982) *J. Exp. Biol.* **98,** 317–327.
8. Atema, J. (1977) in *Olfaction and Taste VI,* eds. Le Magnen, J. & MacLeod, P. (Information Retrieval, London), pp. 165–174.
9. Akers, R. P. (1989) *J. Chem. Ecol.* **15,** 183–208.
10. David, C. T. & Kennedy, J. S. (1987) *Naturwissenschaften* **74,** 194–196.
11. Berg, K., Voigt, R. & Atema, J. (1992) *Biol. Bull. (Woods Hole, Mass.)* **183,** 377–378.
12. Voigt, R. & Atema, J. (1992) *J. Comp. Physiol. A* **171,** 673–683.
13. Voigt, R. & Atema, J. (1994) in *Olfaction and Taste XI,* eds. Kurihara, K., Suzuki, N. & Ogawa, H. (Springer, New York), p. 787.
14. Gomez, G., Voigt, R. & Atema, J. (1994) *J. Comp. Physiol. A* **174,** 803–811.
15. Gomez, G., Voigt, R. & Atema, J. (1994) in *Olfaction and Taste XI,* eds. Kurihara, K., Suzuki, N. & Ogawa, H. (Springer, New York), pp. 788–789.
16. Marion-Poll, F. & Tobin, T. R. (1992) *J. Comp. Physiol. A* **171,** 505–512.
17. Rumbo, E. & Kaissling, K.-E. (1989) *J. Comp. Physiol. A* **165,** 281–291.
18. Christensen, T. & Hildebrand, J. (1988) *Chem. Senses* **13,** 123–130.
19. Mountain, D. & Atema, J. (1993) *Neurosci. Abstr.* **19,** 121.
20. Consi, T. R., Atema, J., Goudey, C. A., Cho, J. & Chryssostomidis, C. (1994) *Proc. IEEE Symp. Auton. Underwater Vehicle Tech.,* 450–455.
21. Bossert, W. H. (1968) *J. Theoret. Biol.* **18,** 157–170.
22. Connor, W. E., Eisner, T., Vander Meer, R. K., Guerrero, A., Ghiringelli, D. & Meinwald, J. (1980) *Behav. Ecol. Sociobiol.* **7,** 55–63.
23. Bushmann, P. & Atema, J. (1993) *Biol. Bull. (Woods Hole, Mass.)* **185,** 319–320.
24. Harrison, F. W. & Humes, A. G., eds. (1992) *Microscopic Anatomy of Invertebrates* (Wiley, New York), Vol. 10.
25. Bushmann, P. & Atema, J. (1994) *Chem. Senses* **19,** 448.
26. Lavalli, K. & Barshaw, D. (1989) *Mar. Behav. Physiol.* **15,** 255–264.
27. Cowan, D. F. & Atema, J. (1990) *Anim. Behav.* **39,** 1199–1206.
28. Breithaupt, T. & Atema, J. (1993) *Biol. Bull. (Woods Hole, Mass.)* **185,** 318.
29. Breithaupt, T., Karavanich, C. & Atema, J. (1994) *Chem. Senses* **19,** 446–447.
30. Karavanich, C. & Atema, J. (1991) *Biol. Bull. (Woods Hole, Mass.)* **181,** 359–360.
31. Atema, J., Jacobson, S., Karnofsky, E., Oleszko-Szuts, A. & Stein, L. (1979) *Mar. Behav. Physiol.* **6,** 227–296.
32. Atema, J., Borroni, P., Johnson, B. R., Voigt, R. & Handrich, L. S. (1989) in *Perception of Complex Smells and Tastes*, eds. Laing, D. L., Cain, W., McBride, R. & Ache, B. W. (Academic, Sydney), pp. 83–100.
33. Atema, J. & Engstrom, D. G. (1971) *Nature (London)* **232,** 261–263.
34. McLeese, D. W. (1973) *J. Fish. Res. Board Can.* **30,** 775–778.
35. Atema, J. (1986) *Can. J. Fish. Aquat. Sci.* **43,** 2283–2290.
36. Wilson, E. O. & Bossert, W. H. (1963) *Recent Prog. Horm. Res.* **19,** 673–716.

Analysis of Chemical Signals by Nervous Systems

JOHN G. HILDEBRAND

All creatures detect and react to chemicals in the external environment. In metazoans possessing a differentiated nervous system, an important function of that system is to detect, analyze, integrate, and generate responses to chemicals in the environment. Among the substances to which these organisms must respond are chemical signals, including pheromones and kairomones. Pheromones are chemical messengers between individuals of the same species, such as the sex attractants of moths and the alarm pheromone of honey bees. Kairomones are chemical messengers between species and adaptively favorable to the recipient, such as attractants and stimulants for insect oviposition and feeding emitted by a host plant. The importance of such chemical signals for survival and reproductive success is reflected in remarkable chemosensory capacities and specializations in diverse species of animals.

After considering the evolutionary origins of the olfactory system and some basic principles of olfaction, this brief review examines one of the most extensively studied examples of neural processing of semiochemical information: the sex pheromone-specific olfactory subsystem in male moths. This male-specific subsystem can be viewed as representing an exaggeration of organizational principles and functional mechanisms that are characteristic of olfactory systems in general.

John Hildebrand is Regents' Professor and director of the ARL Division of Neurobiology at the University of Arizona, Tucson.

ORIGINS OF OLFACTORY SYSTEMS

Consideration of the origins and evolution of chemosensation can help one to begin to understand the themes of olfaction and chemical communication that are common to diverse phyletic groups. This section emphasizes ideas that were propounded with characteristic clarity and elegance by the late Vincent Dethier in his 1990 R. H. Wright Lectures on Olfaction. Because the published version of those lectures (1) may not be widely accessible, some of Dethier's main points are restated here.

Origins of Chemoreception

The universal chemoreceptive capacity of living organisms surely must have arisen in the earliest cells, at the dawn of life billions of years ago (1). That capacity enables a cell to respond to substances without the necessity of internalizing or metabolizing them and is fundamental to the living state.

Studies of unicellular organisms have afforded insights about the origins of chemosensory processes and mechanisms exhibited by metazoa. Thus, modern bacteria such as *Escherichia coli* (2, 3) sense and respond to chemicals in ways that probably resemble those of ancient prokaryotes. *E. coli* possess finely "tuned" receptors for specific substances in the environment, mechanisms for transducing the stimuli and for decoding, integrating, and transmitting information about them, and means to generate appropriate behavioral responses. The motifs of chemoreception are conserved and elaborated in the protists and especially the slime molds (1, 4), suggesting subsequent evolutionary transitions. Dethier observed (1):

> With the advent of multicellularity many cells lost some of their ancient skills, but the organism's capability of sensing the chemical richness of the world was not impaired. Chemoreception became the prime function of specialized strategically situated cells anchored in epithelial sheets. The coupling of chemoreception to motility, that is, to behavioral responses, was accomplished by close association with transmitting systems. Transitional stages between the two functional levels, self-contained unicellular systems and neurally-linked multicellular systems, are preserved in contemporary coelenterates. Here are to be found the earliest metazoan chemoreceptors. In the further evolution of the nervous system there was a division of labor associated with a diversification of kinds of neurons, segregation in which like units gathered together, and compartmentalization of functional assemblies within ganglia.

Whence Olfaction?

Evolution of metazoan chemoreception eventually gave rise to anatomically and functionally distinct chemosensory systems—olfactory and gustatory—which are distinguishable, in organisms that have a central nervous system (CNS), on the basis of the disposition of chemoreceptor cells and the central organization of their afferent axons (1). Whereas both olfaction and taste are served by receptor cells densely arrayed in epithelia and exposed to the environment, the typifying feature of an olfactory system is the projection of axons of olfactory receptor cells (ORCs) to discrete, condensed synaptic glomeruli in the CNS. The difference between the central organization of projections of gustatory receptors and ORCs reflects basic functional differences between these two chemical senses: the numbers of receptor cells, substances that normally stimulate those receptors, and qualities or categories of stimuli that can be discriminated are smaller for taste than for olfaction.

Olfactory glomeruli must have evolved early, because these characteristic structures are present in the "olfactory brains" of modern representatives of ancient marine groups including molluscs (5) and crustaceans (6). Likewise the lampreys, which are extant representatives of the most primitive vertebrates, have relatively large olfactory bulbs with glomeruli and conspicuous mitral cells not unlike those of more advanced vertebrates (7).

As animals emerged from the seas to inhabit the land, chemosensory systems had to adapt to terrestrial conditions. In particular, the olfactory system had to detect sparse molecules of diverse volatile substances in the desiccating ambient atmosphere. In both vertebrates and invertebrates, olfactory organs of marine forms became adapted for smelling on land. For example, antennae, which had appeared in many classes of marine arthropods starting as early as the Cambrian period and can be assumed (on the basis of knowledge about contemporary crustaceans) to have served an "olfactory" function, were brought along—with appropriate modifications—by animals that made the transition to the land. Contemporary representatives of the phylum Onychophora, terrestrial animals that have similarities to both annelids and arthropods and have changed little since the Cambrian period, possess antennae and antennal lobes reportedly containing glomeruli (8), and similar olfactory apparatus is nearly universal among the insects.

The remarkable similarity of glomerular organization in the first-order central olfactory neuropils of essentially all invertebrates and vertebrates that have a differentiated olfactory system has been noted often (e.g., refs. 9–12). Indeed, Dethier (1) argued persuasively that olfactory systems similar to those of contemporary insects and vertebrates, with comparable glomerular organization, were probably already in place 500 million years ago. Referring to environmental odor substances produced, first by photosynthesis and later by mankind's organic chemistry, in varieties far exceeding what could have been "anticipated" by evolving olfactory systems, Dethier observed (1):

> The ability of olfactory systems to cope with this plenitude of stimuli together with the fact that specific volatile compounds became associated with different plants and animals and different body sites, glands, and metabolites, provided exquisitely sensitive and accurate cues to the identities of places, trails, individuals, prey, predators, mates, social groups, and food. Olfaction permitted the development of a heretofore unparalleled perceptual talent.

The importance of that "talent" is evident in the facts that behavior mediated by olfaction is often concerned with intraspecific and interspecific communication or recognition and that olfactory input can have profound effects on the behavioral state of the animal (11).

Comparative and phylogenetic considerations such as those outlined in the preceding paragraphs readily lead to speculation that the olfactory systems of modern animals share common antecedents and therefore probably also share common principles of functional organization and information processing. We might ask, What attributes of chemical stimuli do olfactory systems analyze and encode? How are those features mapped in "neural space" at various levels of the olfactory pathway? How are cells of the pathway organized, and what mechanisms do they use, to accomplish this analysis of odors in the environment?

FROM STIMULANT MOLECULE TO MOLECULAR IMAGES IN THE OLFACTORY SYSTEM

The breadth, precision, and behavioral significance of olfaction result from both peripheral and central mechanisms. The ability of olfactory systems to distinguish myriad odors depends on the response characteristics of ORCs, and hence ultimately on the cascades of molecular and cellular events, leading from molecular recognition at the receptor site to the generation of action potentials in temporal (in each ORC axon) and spatial (across the array of ORC axons) patterns that represent features of the stimulus. That ability also depends on neural circuitry in the CNS through which afferent olfactory information is integrated, abstracted, and recognized. To begin to understand how chemical signals are analyzed by the olfactory system and ultimately affect behavior, we must consider the anatomical and functional organization of the olfactory pathway and the processing performed, and abstraction accomplished, by neural circuitry at each level in that pathway.

Again this review emphasizes stimulating ideas put forth in R. H. Wright Lectures, in this case those of Gordon Shepherd (13). One of his key points is that an understanding of the neurobiology—the organization and function of neural circuits—of the olfactory system is crucial for relating properties of an olfactory stimulus to an animal's behavioral responses to it. Shepherd focused on vertebrate olfaction, but here we consider some aspects of his conceptual framework that appear to have much wider phyletic application. According to this model, olfactory information processing involves generation of a sequence of activity maps, termed "molecular images," in the olfactory pathway. Most of the mechanisms involved in this sequence are adaptations of, and bear similarities to, those underlying vision, immune responses, hormonal communication, chemotaxis of motile unicellular organisms, and other biological processes.

Common Features of Organization of Olfactory Systems

Olfactory systems of diverse vertebrates and invertebrates have certain general organizational and functional features in common. The pathway begins with ORCs residing in an epithelium and interspersed with supporting cells. Ciliary processes of ORC dendrites are relatively exposed, with only aqueous perireceptor fluid [e.g., mucus in vertebrates and sensillum liquor in insects (14)] separating the dendritic membrane from environmental chemicals. Odor molecules must traverse that aqueous phase, perhaps carried by odorant-binding proteins (14), to reach receptor sites in the ciliary dendrites of ORCs. Sensory transduction is believed to be initiated by binding of odor molecules to ORC membrane receptor proteins that are coupled to guanine nucleotide-binding proteins, triggering concatenated events involving multiple intracellular second messengers that ultimately open ion channels and thus generate a receptor potential in the ORC (15). The receptor potential spreads through the dendrite toward the cell body of the ORC and sets up a discharge of action potentials that propagate along the ORC axon to the CNS.

At their entry into the first-order olfactory center in the CNS (e.g., the olfactory bulb in vertebrates and the antennal lobe in insects), fascicles of ORC axons intermingle, and the axons defasciculate and refasciculate before terminating in glomeruli. This regrouping of primary-afferent fibers as they approach their central targets apparently accomplishes a reorganization of the axons, from grouping based on somatotopy to grouping based on odotopy. Each glomerulus encloses arborizations of central neurons that can be classified into numerous types but generally fall into two main classes: projection (or principal or output) neurons that extend axons to subsequent way stations in the pathway and local interneurons confined to the olfactory bulb or lobe. Each ORC axon projects to one glomerulus, and many ORC axons converge on each glomerulus, where they have synaptic connections with neurites of particular types of central neurons (16, 17). On the basis of quantitative estimations performed in a number of vertebrate and invertebrate species (18), the ranges of orders of magnitude of elements associated with the first-order olfactory center appear to be 10^5–10^8 ORC axons projecting into an array of 10–1000 glomeruli, from each of which 1–100 projection neurons relay synaptically processed information about odor stimuli to higher centers in the brain.

Examples of projection neurons are the mitral and tufted cells of the vertebrate olfactory bulb, each with its principal dendrite confined to one glomerulus, and the uniglomerular and multiglomerular projection neurons of the insect antennal lobe. These neurons convey information about odor stimuli, as patterns of action potentials shaped through intra- and interglomerular synaptic circuitry involving various types of local interneurons as well as the projection neurons themselves, to higher order olfactory foci in the CNS (e.g., olfactory cortex in vertebrates, mushroom bodies and lateral protocerebrum in insects) (17, 19). Vertebrate olfactory cortex exhibits circuit properties that prompted Haberly (20) to suggest that olfactory cortex serves as a content-addressable memory for association of odor stimuli with memory traces of previous odors. In view of the striking parallel fiber arrays that characterize both pyriform cortex (20) and the mushroom bodies of insects (21, 22), it is not unreasonable to imagine a similar role for the latter structures. Moreover, mounting evidence suggests that interneuronal circuitry in other regions of the insect protocerebrum shapes descending premotor neural activity. This olfactorily influenced neural activity ultimately participates in the control of behavioral responses to odor stimuli (such as the characteristic flight patterns of male moths stimulated by female sex pheromone; ref. 23; see below).

The Primitives of Olfaction

To begin to understand how the olfactory system constructs molecular images or maps of information about odor substances in neural space, we must consider what properties—or primitives—of the stimulus molecules are being mapped (13). Among the molecular "determinants" that contribute to the "odogenicity" of an odor substance are molecular properties such as presence of functional groups, geometry (e.g., molecular length, position of functional groups, geometry of double bonds), and connectivity (e.g., number and sizes of rings and branching). The ability of odor stimuli to elicit behavior or evoke perceptions depends on multiple molecular properties of the stimulus molecules, including these individual molecular properties or determinants, population properties of a single odor substance, and appropriate mixtures of different odor substances (13). Also important, in addition to such qualitative properties, are odor intensity (concentration of odor substances) and intermittency. A requirement for interruption or intermittency of stimulation for effective olfactory function is found in diverse animals, from crustaceans and insects to air-breathing vertebrates (1).

Molecular Images in Olfactory Pathways

Pioneering efforts to understand the nature of olfactory coding were reported by Adrian (24–27). His work introduced the ideas that different odors activate ORCs in different regions of the olfactory epithelium and that spatiotemporal patterns of ORC firing would suffice to encode different odors. Subsequent studies by many investigators and involving various recording methods (reviewed in refs. 13 and 28) led to the conclusion that, at various levels of the pathway, the olfactory system uses distributed neural activity to encode information about olfactory stimuli.

Different odor substances stimulate different patterns of ORCs in the olfactory epithelium, owing to the different sensitivity spectra of the ORCs (28). The pattern of activity in the epithelium evoked by a particular odor substance constitutes the first molecular image of that stimulus, which represents the determinants of the stimulating molecules (13). Thus, although olfaction is not a spatial sensory modality, in contrast, for example, to vision and somatosensation, the initial representation of an odor stimulus in the olfactory pathway does have spatial structure.

At subsequent levels in the olfactory pathway, new molecular images of the odor stimulus are formed as patterns of activity across an array of neural elements. For example, in the olfactory nerve, which carries ORC axons to the olfactory bulb or lobe of the brain, the pattern of activity across the array of fasciculated primary-afferent fibers constitutes the second molecular image of the stimulus. In the olfactory bulb or lobe, another molecular image takes shape as the pattern of activity across the array of glomeruli, and yet another molecular image is generated as the pattern of activity across the array of projection-neuron axons emanating from the glomeruli. Each molecular image of a particular odor stimulus exemplifies the way neural space is used at that level in the pathway to represent information about the stimulus.

An important insight from many studies (28) is that the response patterns—the molecular images—at various levels in the central olfactory pathway are set up by the differential responses of the ORCs in the peripheral receptor epithelium. These studies also suggest that functional modules, which may correspond to recognizable structural units such as individual glomeruli with their associated cells, in the olfactory bulb or lobe participate in the analysis of olfactory information conveyed to them by primary-afferent ORC axons (28). A characteristic set or pattern of modules would be activated by a given odor stimulus, and particular modules could be shared by the patterns activated by different odor stimuli if the molecular determinants of the stimuli overlap. Thus, for example, experiments using the radioisotopic 2-deoxyglucose method of activity labeling indicated that different odors evoke activity in glomeruli localized in different regions of the olfactory bulb (29). Moreover, recent investigations of the response specificities or "molecular receptive ranges" of individual uniglomerular output neurons (mitral and tufted cells) in the vertebrate olfactory bulb strongly support the idea that the glomeruli are functional units (30–33).

A CASE IN POINT: NEURAL PROCESSING OF SEX-PHEROMONAL INFORMATION IN MOTHS

In many species of insects, olfaction is decisive for the control of several kinds of behavior. Orientation and movement toward, and interactions with, potential mates, appropriate sites for oviposition, sources of food, and hosts for parasitism often involve olfactory signals that initiate, sustain, and guide the behaviors. Because of their prominence in the zoosphere, their economic and medical importance, and their usefulness as models for both behavioral and neurobiological research, insects have been studied extensively to elucidate mechanisms of olfactory control of behavior. Insects respond to a variety of semiochemicals, including pheromones and kairomones. Studies of the responses of insects to such biologically significant odors have shown that the quality and quantity of odor substances in complex mixtures present in the environment are encoded in patterns of activity in multiple ORCs in the antennae. These "messages" are decoded and integrated in the olfactory centers of the CNS and ultimately lead to olfactorily induced changes in the behavior of the insect.

Of paramount interest, both historically and currently, is the attraction of a mating partner by means of a chemical signal—the sex pheromone—released by a receptive individual of one sex and detected by conspecifics of the opposite sex. In moths, these chemical signals are the primary means by which females broadcast their sexual receptiveness over relatively long distances to conspecific males. The male moths respond to the sex-pheromonal signal with well-characterized mate-seeking behaviors involving arousal, patterned upwind flight, short-range orientation to the calling female, and mating (34, 35).

Building upon the work of others (36–39) and paralleling current research in other laboratories on different insect species, we study the olfactory system of the experimentally favorable giant sphinx moth

Manduca sexta. The brief review presented here focuses on the functional organization and physiology of a sexually dimorphic olfactory subsystem in this species [also reviewed elsewhere (19, 23, 40–42)].

The principal long-term goals of this line of research are to understand the neurobiological mechanisms through which the conspecific females' sex pheromone is detected and information about it is integrated with inputs of other modalities in the male moth's brain and to unravel how the message ultimately initiates and controls his characteristic behavioral responses. Pursuit of these goals promises to teach us much about how the brain processes olfactory information and uses it to shape behavior. Our studies to date have persuaded us that the male's olfactory system consists of two parallel subsystems: one is a complex, sexually isomorphic pathway that processes information about plant (and probably other environmental) odors encoded in "across-fiber" patterns of physiological activity and bears striking similarities to the main olfactory pathway in vertebrates, and the other is a sexually dimorphic "labeled-line" pathway specialized to detect and process information about the sex pheromone.

The Sex-Pheromonal Stimulus

The sex pheromones of moths generally are mixtures of two or more chemical components, typically aldehydes, acetates, alcohols, or hydrocarbons, produced in specialized glands by biosynthesis and modification of fatty acids (34). Often, a species-specific blend of components is the message, and males of many moth species, including *M. sexta*, give their characteristic, qualitatively and quantitatively optimal behavioral responses only when stimulated by the correct blend of sex-pheromone components and not by individual components or partial blends lacking key components (43, 44).

Solvent washes of the pheromone gland of female *M. sexta* yield eight C_{16} aldehydes (as well as four C_{18} aldehydes believed not to be pheromone components) (44). A synthetic mixture of the C_{16} aldehydes elicits the same behavioral responses in males as does the signal released by a calling female (44, 45). A blend of two of the components, the dienal (*E*,*Z*)-10,12-hexadecadienal and the trienal (*E*,*E*,*Z*)-10,12,14-hexadecatrienal—hereinafter called components A and B, respectively—elicits an apparently normal sequence of male behavior in a wind tunnel, but the individual components are ineffective (44). Field trapping studies have shown that a blend of the eight C_{16} aldehydes is significantly more effective in attracting males than are blends of fewer components (46), suggesting that all eight C_{16} aldehydes play roles in the communication system of *M. sexta*—i.e., that the sex pheromone of this species is composed of those eight C_{16} aldehydes (44, 45).

Our neurophysiological studies have focused on three important properties of the sex-pheromonal signal: its quality (chemical composition of the blend), quantity (concentrations of components), and intermittency [owing to the fact that the pheromone in the plume downwind from the source exists in filaments and blobs of odor-bearing air interspersed with clean air (47, 48)]. Each of these properties of the pheromonal message is important, as the male moth gives his characteristic behavioral responses only when the necessary and sufficient pheromone components A and B are present in the blend (44), when the concentrations and blend proportions of the components fall within acceptable ranges (49), and when the pheromone blend stimulates his antennae intermittently (39, 50). In our studies, we examine how each of these important aspects of the odor stimulus affects the activity of neurons at various levels in the olfactory pathway.

Detection of the Sex Pheromone

The distalmost, third segment of the antenna of adult *M. sexta* is a long, sexually dimorphic flagellum divided into at least 80 annuli bearing numerous sensilla of several types, the great majority of which are olfactory (51, 52). Antennal flagella of both male and female *M. sexta* have many ORCs that respond to volatile substances given off by plants (53) and presumably are involved in host–plant recognition and discrimination. In addition, the flagella of the male moth possess ORCs specialized to detect the individual key components of the female's sex pheromone (53, 54).

A male flagellum has $\approx 3 \times 10^5$ ORCs associated with $\approx 10^5$ recognized sensilla, of which $\approx 40\%$ are male-specific sensory hairs or sensilla trichodea (51, 53–56). Type I trichoid hairs ($\leq 600\ \mu m$ in length) are typical olfactory sensilla, with single walls and pores, and include two ORCs that send their unbranched dendrites through the lumena of the fluid-filled hairs to their tips (51, 55–57). In most of these sensilla, one of the two male-specific ORCs is highly sensitive and specific to pheromone component A, while the second ORC is tuned to component B (54). Pheromone-specific ORCs in moth antennae thus represent information about stimulus quality by means of their specialization as narrowly tuned input channels. Groups of these cells can "follow" intermittent pheromonal stimuli at naturally occurring frequencies (≤ 10 stimuli per sec) (58).

Olfactory Transduction in Pheromone-Specific Receptor Cells

Physiological and biochemical approaches have yielded increasingly detailed information about mechanisms underlying the transduction of odor stimuli into electrophysiological activity in antennal ORCs (59). In the case of *M. sexta*, primary cell cultures of immunocytochemically identifiable, male-specific ORCs have been studied by means of patch-recording and pharmacological techniques (60–63). These cultured ORCs respond to stimulation by pheromone components with opening of nonspecific cation channels that appear not to be directly gated by pheromone receptors but, instead, to depend on intracellular second messengers mobilized via receptor-coupled guanine nucleotide-binding protein(s). The possibility that these cation channels are controlled by one or more intracellular messengers is consistent with findings from biochemical studies on other species (59). It is likely that these cation channels, which are permeable to Na^+ and K^+ as well as Ca^{2+} ions, play an important role in pheromone transduction.

Functional Organization of Central Olfactory Pathways

Axons of antennal ORCs project through the antennal nerve to enter the brain at the level of the ipsilateral antennal lobe (AL) of the deutocerebrum (52). ORC axons project from the flagellum to targets in the AL, but axons from antennal mechanosensory neurons bypass the AL and project instead to an "antennal mechanosensory and motor center" in the deutocerebrum posteroventral (with respect to the body axis of the animal) to the AL (52, 58, 64). In moths and certain other insect groups, sex-pheromonal information is processed in a prominent male-specific neuropil structure in each AL called the macroglomerular complex (MGC) (16, 52, 64, 65).

In *M. sexta* the AL has a central zone of coarse neuropil (largely neurites of AL neurons) surrounded by an orderly array of glomeruli, including 64 $\pm$ 1 spheroidal "ordinary" glomeruli and, in the male, the sexually dimorphic MGC near the entrance of the antennal nerve into the AL (64, 66). Bordering this neuropil are the lateral, medial, and anterior groups of AL neurons, totaling about 1200 cells (19, 65, 67). The ordinary glomeruli, which are condensed neuropil structures 50–100 μm in diameter, contain terminals of sensory axons and dendritic arborizations of AL neurons, as well as primary-afferent synapses and synaptic connections among AL neurons, and they are nearly surrounded by glial cells (65, 68, 69). Each ORC axon from the antenna terminates within a single glomerulus in the ipsilateral AL (64, 68, 89), where it makes chemical synapses with neurites of AL neurons, primarily local interneurons (65, 68, 70, 71).

With very few exceptions, the neurons in the medial, lateral, and anterior cell groups of the AL fall into two main classes (19, 65, 67, 72, 73). Projection neurons (PNs or output neurons) have dendritic arborizations in the AL neuropil and axons that project out of the AL, and local interneurons (LNs) lack axons and have more or less extensive arborizations confined to the AL neuropil. The PNs relay information about odors, synaptically processed and integrated in the AL by neural circuitry involving sensory axons, LNs and PNs, to olfactory foci in the protocerebrum (67, 72). Many PNs have dendritic arborizations confined to single AL glomeruli and axons that project via the inner antennocerebral tract through the ipsilateral protocerebrum, sending branches into the calyces of the ipsilateral mushroom body and terminating in characteristic olfactory foci in the lateral protocerebrum (67, 73, 74). Other PNs have arborizations in one or more AL glomeruli and send axons via different antennocerebral tracts to characteristic regions of the lateral and inferior protocerebrum (19, 67, 73).

Axons of male-specific antennal ORCs specialized to detect components of the sex pheromone project exclusively to the MGC (64, 89), and all AL neurons that respond to antennal stimulation with sex pheromone components have arborizations in the MGC (65, 72, 73). The MGC in *M. sexta* has two major, easily distinguishable divisions: a donut-shaped neuropil structure (the "toroid") and a globular structure (the "cumulus") adjacent to the toroid and closer to the entrance of the antennal nerve into the AL (74). AL PNs that respond to antennal stimulation with sex pheromone component A have arborizations in the toroid and PNs responsive to component B, in the cumulus (74). Thus first-order synaptic processing of sensory information about these key components of the sex pheromone apparently is confined to different, distinctive neuropil regions of the MGC.

Stimulus Quality

By means of intracellular recording and staining methods, we have examined the responses of AL neurons to stimulation of the ipsilateral antenna with each of the sex pheromone components as well as partial and complete blends (75). In accordance with results of behavioral and sensory-receptor studies, components A and B are the most effective and potent sex pheromone components for eliciting physiological responses in the male-specific AL neurons. On the basis of these responses, we classified the neurons into two broad categories: pheromone generalists and pheromone specialists (76). Pheromone generalists are neurons that respond similarly to stimulation of either the component A input channel or the component B input channel and do not respond differently when the complete, natural pheromone blend is presented to the antenna. In contrast, pheromone specialists are neurons that can discriminate between antennal stimulation with component A and stimulation with component B. There are several types of pheromone specialists. Some receive input only from the component A input channel or the component B input channel and thus preserve information about individual components of the blend.

An important subset of pheromone-specialist PNs in male *M. sexta* receives input from both component A and component B input channels, described above, but the physiological effects of the two inputs are opposite (72). That is, if antennal stimulation with component A leads to excitation, then stimulation with component B inhibits the interneuron, and vice versa. Simultaneous stimulation of the antenna with both components A and B elicits a mixed inhibitory and excitatory response in these special PNs. Thus these neurons can discriminate between the two inputs based upon how each affects the spiking activity of the cell. These PNs also respond uniquely to the natural pheromone blend released by the female: these pheromone specialist neurons have enhanced ability to follow intermittent pheromonal stimuli occurring at natural frequencies of $\leq$10 stimuli per sec (77).

Stimulus Quantity

Numerous studies in the field and in wind tunnels have shown that pheromone-mediated orientation is dose-dependent (35). We therefore have examined the ability of AL neurons to encode changes in the concentration of a pheromonal stimulus (76). When a male's antenna is stimulated with a series of pheromonal stimuli of graded concentrations, MGC PNs exhibit various dose–response relationships. In some of these PNs, the dynamic range of the cell extends up to the highest concentration tested [0.1 female equivalent (FE) of sex pheromone component(s) in the odor-delivery source], but in other MGC PNs, inactivation of spiking occurs between 0.01 and 0.05 FE. Some PNs that have this ability to encode quantitative information about the pheromone yield a dose–response relationship, measured in terms of the number of spikes elicited, that is quite linear up to 0.05 FE but falls off above this concentration. The maximum instantaneous frequency of spiking, however, continues to increase up to the highest concentration tested (0.1 FE). A corresponding increase in the amplitude of the membrane depolarization can also be seen.

Stimulus Intermittency

A third important characteristic of a female moth's sex-pheromone plume is its nonuniformity. Simulation of odor plumes using ionized air has shown clearly that a plume is not a simple concentration gradient but instead is distinctly filamentous and discontinuous (47, 48). Furthermore, abundant behavioral evidence shows that a male moth's ability to locate a pheromone source is greatly improved if the odor plume is discontinuous (35). Because spatial discontinuity of the pheromonal signal in the environment is detected by a flying male moth as temporally intermittent stimuli, intermittent pheromonal stimuli received by a male's antenna must be registered by MGC PNs. We discovered that certain pheromone specialist PNs have greatly enhanced ability to follow pulsed pheromonal stimuli with corresponding bursts of impulses. These are the PNs cited above that can discriminate between components A and B because one of these key components excites the cells while the other inhibits them. This inhibitory input to such PNs enhances their ability to follow brief pulses of pheromone blends delivered at frequencies up to 10 stimuli per sec by controlling the duration of excitatory responses and preparing the PN for the next bout of excitation (77).

Synaptic Mechanisms in the AL

Having characterized many AL neurons both morphologically and physiologically, we have sought to explain how their characteristic patterns of responses to olfactory stimuli are generated. To accomplish this mechanistic goal, we must analyze the synaptic "wiring" of the AL, test physiologically for synaptic interactions between known types of AL neurons, identify neurotransmitters and synaptic mechanisms employed by AL neurons for intercellular communication, and seek evidence for and mechanisms of integration of other modalities with olfactory inputs in the ALs.

Synapses between ORC axons and their AL target neurons are excitatory and appear to be mediated by the neurotransmitter acetylcholine acting through nicotinic cholinoceptive mechanisms (40, 71, 78–82). Another prominent neurotransmitter in the ALs is γ-aminobutyric acid (GABA) (79, 83). GABA immunocytochemistry has revealed that all of the GABA-immunoreactive neurons in the AL have somata in the large lateral cell group of the AL (84). There are ≈350 GABA-immunoreactive LNs and 110 GABA-immunoreactive PNs (i.e., ≈30% of the neurons in the lateral cell group may be GABAergic) (19, 84). Most (and possibly all) of the LNs are GABA-immunoreactive and thus may be inhibitory interneurons. The important inhibitory postsynaptic potential (IPSP) that enables certain pheromone specialist MGC PNs to follow intermittent pheromonal stimuli (see above) appears to be due to chemical-synaptic transmission mediated by GABA (85). This IPSP reverses below the PNs' resting potential and is mediated by an increased Cl^- conductance. This IPSP can be inhibited reversibly by picrotoxin, which blocks GABA receptor-gated Cl^- channels, and by bicuculline, a blocker of vertebrate

$GABA_A$ receptors. Furthermore, applied GABA hyperpolarizes the postsynaptic neuron, and this response can be blocked reversibly by bicuculline, indicating that bicuculline directly blocks the GABA receptors. Such GABAergic synaptic transmission is essential to the enhanced ability of the specialized AL PNs to follow intermittent pheromonal stimuli (77).

Synaptic Interactions Between AL Neurons

We have tested the idea that this inhibition of PNs is mediated through LNs by directly recording synaptic interactions between pairs of AL neurons (70). Current was passed into one neuron while the postsynaptic activity in the other neuron was monitored. None of the PN–PN pairs examined showed any current-induced interactions, but a significant proportion of the LN–PN pairs studied exhibited such interactions, all of which were unidirectional. That is, LN activity could influence PN activity, but not vice versa. Depolarizing current injected into an LN, causing it to produce spikes, was associated with cessation of firing in the PN. Spike-triggered averaging revealed a weak, prolonged IPSP in the PN. Cross-correlation analysis also revealed a weak inhibitory interaction, polarized in the direction from LN to PN. When the simultaneous responses of the two neurons to olfactory stimulation of the ipsilateral antenna with pheromone component A were recorded, a brief period of inhibition was observed in the LN, and this was followed shortly thereafter by a transient increase in the firing frequency of the PN. This suggests that the "excitation" of the PN is due to disinhibition. The LN that synaptically inhibits the PN thus may itself be inhibited by olfactory inputs, probably through another LN.

Higher Order Processing of Pheromonal Information in the CNS

After synaptic processing in the AL, information about sex pheromone and other odors is relayed to higher centers in the protocerebrum by way of the axons of PNs with arborizations in the MGC. Toward the goal of understanding how pheromonal information controls the behavior of male moths, we have begun to explore the physiological and morphological properties of neurons in the protocerebrum that respond to stimulation of the antennae with sex pheromone or its components (86, 87).

Many pheromone-responsive protocerebral neurons have arborizations in the lateral accessory lobes (LALs), which are situated lateral to the central body on each side of the protocerebrum and appear to be important for processing of olfactory information (86). Each LAL is linked, by neurons with arborizations in it, to the ipsilateral superior protocerebrum as well as the lateral protocerebrum, where axons of AL

PNs terminate (67, 72, 73). The LALs are also linked to each other by bilateral neurons with arborizations in each LAL. Neuropil adjacent to the LAL contains branches of many neurons that descend in the ventral nerve cord. Local neurons link the LAL to this adjacent neuropil. Some descending neurons also have arborizations in the LAL. Thus, the LAL is interposed in the pathway of olfactory information flow from the AL through the lateral protocerebrum to descending neurons.

All protocerebral neurons observed to date to respond to antennal stimulation with pheromone were excited. Although brief IPSPs were sometimes elicited in mixed inhibitory/excitatory responses, sustained inhibition was not observed. Certain protocerebral neurons show long-lasting excitation (LLE) that sometimes outlasts the olfactory stimuli by up to 30 sec. In some other protocerebral neurons, pheromonal stimuli elicit brief excitations that recover to background firing rates <1 sec after stimulation. LLE is more frequently elicited by the complete sex pheromone blend or a mixture of components A and B than by either component alone. LLE in response to pheromonal stimuli was observed in $>50\%$ of the bilateral protocerebral neurons sampled that had arborizations in the LALs. Fewer than 10% of the protocerebral local neurons examined exhibited LLE in response to similar stimuli. In *M. sexta*, AL PNs responding to pheromone components do not show LLE (72, 73). Thus, LLE appears not to be produced at early stages of olfactory processing in the AL, but first occurs at the level of the protocerebrum.

These findings suggest that the LAL is an important region of convergence of neurons from olfactory foci elsewhere in the protocerebrum. Synaptic interactions in the LAL may mediate integration of both ipsilateral and bilateral olfactory information prior to its transmission to the bilateral pool of descending neurons. LLE appears to be one important kind of physiological response that may be transmitted to thoracic motor centers. How this LLE might contribute to the generation of the male moth's characteristic behavioral responses to sex pheromone is currently unknown.

CONCLUSION

The male moth's pheromone-analyzing olfactory subsystem is composed of pheromone-specific antennal ORCs projecting to the similarly specialized, anatomically defined MGC in the AL and MGC output neurons that project to olfactory foci in the protocerebrum. This subsystem is an example of a labeled-line pathway (18). Its specialization to detect, amplify, and analyze features of sex-pheromonal signals and its consequent exaggeration of common olfactory organizational principles and functional mechanisms make this system especially favorable for experimentation as well as for computational modeling (88).

In this specialized subsystem, the molecular images attributable to at least the first several levels of the pathway seem, to a first approximation, to be relatively simple. For components A and B of the pheromone, the first molecular image at the level of the antennal olfactory epithelium would be a pattern of activity corresponding to the orderly anatomical pattern of distribution of type I trichoid sensilla on each annulus of the flagellum. At the next level, the molecular image would be a pattern of activity on a cross-section of the antennal nerve corresponding to the pattern of fasciculation of axons of ORCs specialized to respond to components A and B. At the level of the glomeruli of the AL, the molecular image would map isomorphically with the major subdivisions of the MGC, the component A-specific toroid and the component B-specific cumulus, and so forth through the MGC output tracts and protocerebral olfactory foci to descending premotor pathways, one can envision hypothetical molecular images of the pheromone at each successive level.

This way of viewing the olfactory system spotlights the kinds of information about odor stimuli to which the brain attends. For example, although the blend of components is essential to evoke and sustain the normal male responses to the sex pheromone, information about specific components is preserved through many levels of the pathway. Thus it appears that information about single components as well as their blend may be important for chemical communication in these insects.

SUMMARY

Intraspecific and interspecific communication and recognition depend on olfaction in widely diverse species of animals. Olfaction, an ancient sensory modality, is based on principles of neural organization and function that appear to be remarkably similar throughout the zoosphere. Thus, the "primitives" of olfactory stimuli that determine the input information of olfaction, the kinds of "molecular images" formed at various levels in the olfactory pathway, and the cellular mechanisms that underlie olfactory information processing are comparable in invertebrates and vertebrates alike. A case in point is the male-specific olfactory subsystem in moths, which is specialized to detect and analyze the qualitative, quantitative, and temporal features of the conspecific females' sex-pheromonal chemical signal. This olfactory subsystem can be viewed, and is here presented, as a model in which common principles of organization and function of olfactory systems in general are exaggerated to serve the requirements of a chemical communication system that is crucial for reproductive success.

This article is dedicated to the memory of Vincent G. Dethier. I thank Drs. Thomas A. Christensen and Leslie P. Tolbert for helpful comments on the manuscript and my many present and former coworkers and collaborators, who contributed to the work from my laboratory reviewed in this paper. Our research in this area has been supported by grants from the National Institutes of Health, National Science Foundation, and the Department of Agriculture and is currently supported by National Institutes of Health Grants AI-23253, NS-28495, and DC-00348.

REFERENCES

1. Dethier, V. G. (1990) in *Frank Allison Linville's R. H. Wright Lectures on Olfactory Research*, ed. Colbow, K. (Simon Fraser Univ., Burnaby, BC, Canada) pp. 1–37.
2. Adler, J. (1987) *Ann. N.Y. Acad. Sci.* **510,** 95–97.
3. Adler, J. (1990) in *Frank Allison Linville's R. H. Wright Lectures on Olfactory Research*, ed. Colbow, K. (Simon Fraser Univ., Burnaby, BC, Canada), pp. 38–60.
4. Van Houten, J. & Preston, R. R. (1987) *Ann. N.Y. Acad. Sci.* **510,** 16–22.
5. Chase, R. (1986) *Cell Tissue Res.* **246,** 567–273.
6. Nässel, D. R. & Elofsson, R. (1987) in *Arthropod Brain*, ed. Gupta, A. P. (Wiley, New York), pp. 111–133.
7. Healey, E. G. (1972) in *The Biology of Lampreys*, eds. Hardisty, M. W. & Potter, I. C. (Academic, New York), Vol. 2, pp. 307–372.
8. Schürmann, F.-W. (1987) in *Arthropod Brain*, ed. Gupta, A. P. (Wiley, New York), pp. 159–180.
9. Belloni, G. (1883) *Arch. Ital. Biol.* **3,** 191–196.
10. Hanström, B. (1928) *Vergleichende Anatomie des Nervensystems der Wirbellosen Tiere* (Springer, Berlin).
11. Maynard, D. M. (1967) in *Invertebrate Nervous Systems*, ed. Wiersma, C. A. G. (Univ. of Chicago Press, Chicago), pp. 231–255.
12. Shepherd, G. M. (1972) *Physiol. Rev.* **52,** 864–917.
13. Shepherd, G. M. (1990) in *Frank Allison Linville's R. H. Wright Lectures on Olfactory Research*, ed. Colbow, K. (Simon Fraser Univ., Burnaby, BC, Canada), pp. 61–109.
14. Carr, W. E. S., Gleeson, R. A. & Trapido-Rosenthal, H. G. (1990) *Trends Neurosci.* **13,** 212–215.
15. Lancet, D., Lazard, D., Heldman, J., Khen, M. & Nef, P. (1988) *Cold Spring Harbor Symp. Quant. Biol.* **53,** 343–348.
16. Boeckh, J. & Tolbert, L. P. (1993) *Microsc. Res. Tech.* **24,** 260–280.
17. Shepherd, G. M. & Greer, C. A. (1990) in *The Synaptic Organization of the Brain*, ed. Shepherd, G. M. (Oxford Univ. Press, Oxford), pp. 133–169.
18. Boeckh, J., Distler, P., Ernst, K. D., Hösl, M. & Malun, D. (1990) in *Chemosensory Information Processing*, ed. Schild, D. (Verlag, Berlin), pp. 201–227.
19. Homberg, U., Christensen, T. A. & Hildebrand, J. G. (1989) *Annu. Rev. Entomol.* **34,** 477–501.
20. Haberly, L. B. (1985) *Chem. Senses* **10,** 219–238.
21. Schürmann, F.-W. (1987) in *Arthropod Brain*, ed. Gupta, A. P. (Wiley, New York), pp. 231–264.
22. Erber, J., Homberg, U. & Gronenberg, W. (1987) in *Arthropod Brain*, ed. Gupta, A. P. (Wiley, New York), pp. 485–511.

23. Hildebrand, J. G., Christensen, T. A., Arbas, E. A., Hayashi, J. H., Homberg, U., Kanzaki, R. & Stengl, M. (1992) in *Proceedings of NEUROTOX 91—Molecular Basis of Drug and Pesticide Action*, ed. Duce, I. R. (Elsevier, London), pp. 323–338.
24. Adrian, E. D. (1950) *Br. Med. Bull.* **6,** 330–331.
25. Adrian, E. D. (1951) *L'Annee Psychol.* **50,** 107–113.
26. Adrian, E. D. (1953) *Acta Physiol. Scand.* **29,** 5–14.
27. Adrian, E. D. (1963) in *Olfaction and Taste I*, ed. Zotterman, Y. (Pergamon, London), pp. 1–4.
28. Kauer, J. S. & Cinelli, A. R. (1993) *Microsc. Res. Tech.* **24,** 157–167.
29. Stewart, W. B., Kauer, J. S. & Shepherd, G. M. (1979) *J. Comp. Neurol.* **185,** 715–734.
30. Buonviso, N. & Chaput, M. A. (1990) *J. Neurophysiol.* **63,** 447–454.
31. Mori, K., Mataga, N. & Imamura, K. (1992) *J. Neurophysiol.* **67,** 786–789.
32. Imamura, K., Mataga, N. & Mori, K. (1992) *J. Neurophysiol.* **68,** 1986–2002.
33. Katoh, K., Koshimoto, H., Tani, A. & Mori, K. (1993) *J. Neurophysiol.* **70,** 2161–2175.
34. Birch, M. C. & Haynes, K. F. (1982) *Insect Pheromones* (Arnold, London).
35. Baker, T. C. (1989) *Experientia* **45,** 248–262.
36. Boeckh, J. & Boeckh, V. (1979) *J. Comp. Physiol.* **132,** 235–242.
37. Boeckh, J., Ernst, K.-D., Sass, H. & Waldow, U. (1984) *J. Insect Physiol.* **30,** 15–26.
38. Kaissling, K.-E. (1987) *R. H. Wright Lectures on Insect Olfaction*, ed. Colbow, K. (Simon Fraser Univ., Burnaby, BC, Canada).
39. Kaissling, K.-E. (1990) *Verh. Dtsch. Zool. Ges.* **83,** 109–131.
40. Christensen, T. A. & Hildebrand, J. G. (1987) in *Arthropod Brain: Its Evolution, Development, Structure and Functions*, ed. Gupta, A. P. (Wiley, New York), pp. 457–484.
41. Hildebrand, J. G. (1985) in *Model Neural Networks and Behavior*, ed. Selverston, A. I. (Plenum, New York), pp. 129–148.
42. Hildebrand, J. G. & Christensen, T. A. (1994) in *Olfaction and Taste XI: Proceedings of the 11th International Symposium on Olfaction and Taste/27th Japanese Symposium on Taste and Smell*, eds. Kurihara, K., Suzuki, N. & Ogawa, H. (Springer, Heidelberg), in press.
43. Linn, C. E., Jr., & Roelofs, W. L. (1989) *Chem. Senses* **14,** 421–437.
44. Tumlinson, J. H., Brennan, M. M., Doolittle, R. E., Mitchell, E. R., Brabham, A., Mazomenos, B. E., Baumhover, A. H. & Jackson, D. M. (1989) *Arch. Insect Biochem. Physiol.* **10,** 255–271.
45. Starratt, A. N., Dahm, K. H., Allen, N., Hildebrand, J. G., Payne, T. L. & Röller, H. (1979) *Z. Naturforsch.* **34c,** 9–12.
46. Tumlinson, J. H., Mitchell, E. R., Doolittle, R. E. & Jackson, D. M. (1994) *J. Chem. Ecol.* **20,** 579–591.
47. Murlis, J. & Jones, C. D. (1981) *Physiol. Entomol.* **6,** 71–86.
48. Murlis, J., Willis, M. A. & Cardě, R. T. (1990) in *ISOT X: Proceedings of the Tenth International Symposium on Olfaction and Taste*, ed. Døving, K. B. (Univ. of Oslo, Oslo), pp. 6–17.
49. Linn, C. E., Jr., Campbell, M. G. & Roelofs, W. L. (1985) *J. Chem. Ecol.* **12,** 659–668.
50. Baker, T. C. (1990) in *ISOT X: Proceedings of the Tenth International Symposium on Olfaction and Taste*, ed. Døving, K. B. (Univ. of Oslo, Oslo), pp. 18–25.
51. Sanes, J. R. & Hildebrand, J. G. (1976) *Dev. Biol.* **51,** 282–299.
52. Hildebrand, J. G., Matsumoto, S. G., Camzine, S. M., Tolbert, L. P., Blank, S.,

Ferguson, H. & Ecker, V. (1980) *Insect Neurobiology and Pesticide Action (Neurotox 79)* (Soc. Chem. Ind., London), pp. 375–382.
53. Schweitzer, E. S., Sanes, J. R. & Hildebrand, J. G. (1976) *J. Insect Physiol.* **22,** 955–960.
54. Kaissling, K.-E., Hildebrand, J. G. & Tumlinson, J. H. (1989) *Arch. Insect Biochem. Physiol.* **10,** 273–279.
55. Keil, T. A. (1989) *Tissue Cell* **21,** 139–151.
56. Lee, J.-K. & Strausfeld, N. J. (1990) *J. Neurocytol.* **19,** 519–538.
57. Sanes, J. R. & Hildebrand, J. G. (1976) *Dev. Biol.* **51,** 300–319.
58. Marion-Poll, F. & Tobin, T. R. (1992) *J. Comp. Physiol.* **171,** 505–512.
59. Stengl, M., Hatt, H. & Breer, H. (1992) *Annu. Rev. Physiol.* **54,** 665–681.
60. Stengl, M. & Hildebrand, J. G. (1990) *J. Neurosci.* **10,** 837–847.
61. Zufall, F., Stengl, M., Franke, C., Hildebrand, J. G. & Hatt, H. (1991) *J. Neurosci.* **11,** 956–965.
62. Stengl, M., Zufall, F., Hatt, H. & Hildebrand, J. G. (1992) *J. Neurosci.* **12,** 2523–2531.
63. Stengl, M. (1993) *J. Exp. Biol.* **178,** 125–147.
64. Camazine, S. M. & Hildebrand, J. G. (1979) *Soc. Neurosci. Abstr.* **5,** 155.
65. Matsumoto, S. G. & Hildebrand, J. G. (1981) *Proc. R. Soc. London B* **213,** 249–277.
66. Rospars, J. P. & Hildebrand, J. G. (1992) *Cell Tissue Res.* **270,** 205–227.
67. Homberg, U., Montague, R. A. & Hildebrand, J. G. (1988) *Cell Tissue Res.* **254,** 255–281.
68. Tolbert, L. P. & Hildebrand, J. G. (1981) *Proc. R. Soc. London B* **213,** 279–301.
69. Oland, L. A. & Tolbert, L. P. (1987) *J. Comp. Neurol.* **255,** 196–207.
70. Christensen, T. A., Waldrop, B. R., Harrow, I. D. & Hildebrand, J. G. (1993) *J. Comp. Physiol. A* **173,** 385–399.
71. Waldrop, B. & Hildebrand, J. G. (1989) *J. Comp. Physiol. A* **164,** 433–441.
72. Christensen, T. A. & Hildebrand, J. G. (1987) *J. Comp. Physiol. A* **160,** 553–569.
73. Kanzaki, R., Arbas, E. A., Strausfeld, N. J. & Hildebrand, J. G. (1989) *J. Comp. Physiol. A* **165,** 427–453.
74. Hansson, B., Christensen, T. A. & Hildebrand, J. G. (1991) *J. Comp. Neurol.* **312,** 264–278.
75. Christensen, T. A., Hildebrand, J. G., Tumlinson, J. H. & Doolittle, R. E. (1989) *Arch. Insect Biochem. Physiol.* **10,** 281–291.
76. Christensen, T. A. & Hildebrand, J. G. (1990) in *ISOT X: Proceedings of the Tenth International Symposium on Olfaction and Taste,* ed. Døving, K. B. (Univ. of Oslo, Oslo), pp. 18–25.
77. Christensen, T. A. & Hildebrand, J. G. (1988) *Chem. Senses* **13,** 123–130.
78. Sanes, J. R., Prescott, D. J. & Hildebrand, J. G. (1977) *Brain Res.* **119,** 389–402.
79. Maxwell, G. D., Tait, J. F. & Hildebrand, J. G. (1978) *Comp. Biochem. Physiol.* **61C,** 109–119.
80. Sanes, J. R. & Hildebrand, J. G. (1976) *Dev. Biol.* **52,** 105–120.
81. Hildebrand, J. G., Hall, L. M. & Osmond, B. C. (1979) *Proc. Nat. Acad. Sci. USA* **76,** 499–503.
82. Stengl, M., Homberg, U. & Hildebrand, J. G. (1990) *Cell Tissue Res.* **262,** 245–252.
83. Kingan, T. G. & Hildebrand, J. G. (1985) *Insect Biochem.* **15,** 667–675.
84. Hoskins, S. G., Homberg, U., Kingan, T. G., Christensen, T. A. & Hildebrand, J. G. (1986) *Cell Tissue Res.* **244,** 243–252.
85. Waldrop, B., Christensen, T. A. & Hildebrand, J. G. (1987) *J. Comp. Physiol. A* **161,** 23–32.

86. Kanzaki, R., Arbas, E. A. & Hildebrand, J. G. (1991) *J. Comp. Physiol. A* **168,** 281–298.
87. Kanzaki, R., Arbas, E. A. & Hildebrand, J. G. (1991) *J. Comp. Physiol. A* **169,** 1–14.
88. Linster, C., Masson, C., Kerszberg, M., Personnaz, L. & Dreyfus, G. (1993) *Neural Comput.* **5,** 228–241.
89. Christensen, T. A., Harrow, I. D., Cuzzocrea, C., Randolph, P. W. & Hildebrand, J. G. (1995) *Chem. Senses,* in press.

Chemical Ecology: A View from the Pharmaceutical Industry

LYNN HELENA CAPORALE

Effective medical treatments are available for many diseases. Recent discoveries include drugs that treat hypertension, ulcers, and many bacterial infections (for examples, see ref. 1). For other devastating diseases, such as diabetes, cancer, AIDS, stroke, and Alzheimer disease, at best only inadequate therapy is available; new approaches to treat these diseases are sought intensively by researchers within the pharmaceutical industry and elsewhere. In the future, new infectious agents may be expected to emerge as important threats, much as human immunodeficiency virus (HIV) has challenged us. Because of the magnitude of these challenges, we continue to seek to improve the strategies by which we discover new drugs.

Most of the effort to create and discover medicinal molecules occurs in the research laboratories of the pharmaceutical industry. Drug discovery demands a coordinated effort among people with diverse talents. Researchers who elucidate disease mechanisms (which point to macromolecules that would make good drug discovery targets) work with those skilled in isolating or synthesizing new molecules and with those who can determine the safety and effectiveness of these molecules. Pharmaceutical company researchers begin the search for clues to the type of molecules that, for example, inhibit a targeted enzyme, by testing diverse structures in their compound collections. These collections consist of

Lynn Caporale is vice president of strategic development at CombiChem, Inc., La Jolla, California.

compounds synthesized previously in the course of other research programs. Discovery is optimized by scanning as wide a range of structural types as possible (within resource constraints), including compounds found in nature.

Drug discovery has benefited significantly from the study of compounds that organisms have evolved for their own purposes. All of the drugs discovered at the Merck Research Laboratories that became available to patients in the last decade emerged from programs that benefited from knowledge of biological diversity. Some drugs were discovered via natural product screens (2, 3). Others emerged from medicinal chemistry efforts that drew on knowledge such as how the venom of a Brazilian viper lowers blood pressure (4–6). Another program was inspired by a human genetic variation that indicated that a particular enzyme would be a good drug target (7).

SELECTION OF DRUG DISCOVERY TARGETS

A key step in the process of selecting a molecular target for a drug discovery program involves a demonstration that altering the activity of the proposed target should affect the disease. This is illustrated by the choice of the HIV protease as a target. When the HIV genome was sequenced, Asp-Thr-Gly was recognized, based on previous work with other organisms, as a sequence found in aspartyl proteases. Viral-encoded proteases were known to free active proteins from a viral polyprotein precursor. Therefore, it was considered possible that inactivating the proposed HIV aspartyl protease might prevent the formation of viable virus and, thus, that the HIV protease would be an important target for a drug discovery effort (8). To test this hypothesis, a genetic approach was taken to inactivate the aspartyl protease, changing the proposed active site Asp to Asn. This mutation did indeed inactivate the HIV protease; further, mutant HIV, lacking an active protease, was not viable (Figure 1). With this proof of concept in hand, an intensive effort was mounted to find an HIV protease inhibitor. This proof of concept has been reinforced as small molecule inhibitors of the HIV protease have subsequently been found to block the spread of the virus in tissue culture (9).

The search for inhibitors of the HIV protease began with compounds that had been prepared during prior work to inhibit related proteases, notably renin, an aspartyl protease involved in blood pressure regulation (9, 10). While new structures that inhibit the HIV protease have emerged from natural product screens (11, 12) [one structure that inhibits aspartyl proteases, the statine moiety, had been discovered through screening bacterial natural products before the emergence of HIV (10)], the most significant advances in this program have come through an extensive

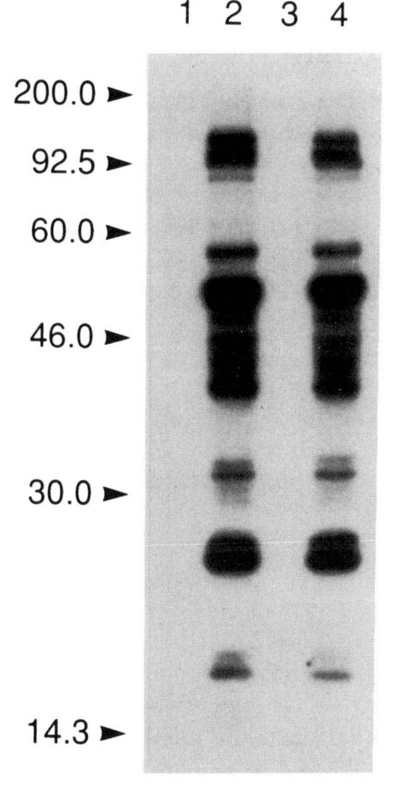

FIGURE 1 Selection of the HIV protease as a drug discovery target. Demonstration that inactivation of the HIV protease interferes with HIV infectivity. Immunoblot of HIV proteins in mock-transfected cells (lane 1), cells transfected with HIV containing active protease (lanes 2 and 4), and cells transfected with HIV in which the protease was inactivated. Reprinted with permission from ref. 8 (copyright Kohl *et al.*).

medicinal chemistry effort that built upon prior work with aspartyl proteases. Inhibitors had to be found that spared other important proteases (such as renin). A more difficult challenge was to discover nontoxic molecules that are bioavailable and have a half-life long enough to be able to block the virus not only in a laboratory assay but also in a person (9, 13, 14).

It has proven extraordinarily difficult to obtain an inhibitor of sufficient potency (15), bioavailability, and safety to be used in the clinic. The ability of HIV to mutate, and thus to develop resistance to many inhibitors, is an additional hurdle. In spite of the difficulties, chemists persisted, motivated by the knowledge that inactivating this HIV-encoded enzyme inactives the virus. Their efforts have provided HIV protease inhibitors that are being evaluated for clinical efficacy (16).

Examples of successful natural product screens directed against proven targets include screens designed to look for inhibitors of bacterial cell wall biosynthesis. These screens were established once it became clear that certain very useful antibiotics of low toxicity inhibit this pathway. This strategy led to the discovery of a variety of potent broad-spectrum antibiotics (17). Advances in analytical methodology facilitated the isolation and characterization of unstable molecules present in minute quantities in source organisms, such as the potent broad-spectrum antibiotic thienamycin (3).

As our understanding of biology deepens, we are increasingly able to select (and obtain in sufficient quantities) a specific protein target for a drug discovery effort (18). Identification of human genes that predispose to diseases of complex etiology, such as Alzheimer disease (19), may provide clues to biochemical sites of intervention. Thus, many future therapies may be targeted to (i.e., discovered by screening against) not the most common human sequence at a locus but disease-associated alleles. Progress in cloning DNA has placed in our hands an array of receptor subtypes and of isozymes. [For example, there are at least 14 human serotonin receptor subtypes (20).] Different receptor subtypes may couple the action of the signaling ligand (which might be a hormone or a neurotransmitter) to different second messenger systems (e.g., activating adenyl cyclase or opening a potassium channel). Subtypes typically have distinct tissue distribution (i.e., one may be found in the brain, another in the heart, or both may be in the same region of the brain, but one may be postsynaptic and the other presynaptic). Consequently, the effects of drugs designed to selectively activate or inhibit one subtype can be confined to a subset of the functions of the endogenous ligand. Many side effects of older drugs are caused by binding to nontarget receptor subtypes. Therefore, selecting one subtype as a target and counterscreening against nontarget subtypes provides a rational approach to broaden the window between efficacy and side effects. Targeting an isozyme provides a similar advantage. For example, the nonselective phosphodiesterase inhibitor theophylline, which is widely used to treat asthma, has undesired cardiovascular and gastrointestinal side effects. A selective inhibitor of one of the phosphodiesterase isozymes might provide an improved treatment for asthma [if theophylline's desired and undesired effects arise from inhibition of different phosphodiesterase isozymes (21)].

It is intriguing to reflect on why there might be more than one receptor subtype acting through the same second messenger. As receptor subtypes with distinct amino acid sequences interact with distinct sets of drugs, are there endogenous molecules (perhaps metabolites of the ligand for which we have named the receptor) to which receptor subtypes are differentially sensitive or might subtypes have evolved to bind to different conformations of the same ligand? A subtype that binds to a low-energy conformation with a higher affinity would sense a lower concentration of the endogenous ligand than that detected by other receptors; the latter would be activated only when ligand concentrations rise. Sensitivity of histamine receptor subtypes to different concentrations of histamine has been proposed to allow triggering of the H1 subtype by low concentrations; higher concentrations, sensed by the H2 subtype, can feed back to shut down an inflammatory response (22). If this is so, once we understand the biological rationale for assigning a given sensitivity to a given subtype, we will be better able to design drugs for a specific purpose. We would select, from the array of possible conformations of the natural ligand, the appropriate conformation to mimic with our drug.

Drug discovery targets can be suggested by the study of organisms that have evolved the ability to regulate our biochemistry. For example, there is much we can learn from a study of infectious organisms. Parasites and viruses often manipulate key regulatory steps in host defenses (23, 24). Pathogenic *Yersinia*, a genus that includes *Yersinia pestis*, which causes bubonic plague, interferes with the cellular immune response of a mammalian host (25). The entry of *Salmonella* into a cell appears to be aided by its use of a leukotriene D_4-activated calcium channel that *Salmonella* activates by triggering host leukotriene D_4 synthesis. *Salmonella* activates leukotriene D_4 synthesis via the epidermal growth factor receptor, which activates the mitogen-activated protein kinase, which in turn activates phospholipase A_2 (25). Some viruses have evolved the ability to override cell death programs that otherwise can eliminate virus-infected cells (26, 27). *Shigella flexneri*, on the other hand, eliminates our first line of defense against invasion by triggering a cell death program in mammalian macrophages (28). If we can learn to manipulate key biochemical control points that trigger or block cell death as well as these pathogens do, then we could save neurons in neurodegenerative disease or kill cells in a tumor.

Cyclosporin (originally discovered due to its antifungal activity) and, more recently, FK506 are used clinically to protect against transplant rejection. Elucidation of the biochemical mechanism of action of these microorganism-derived immunosuppressants [complexes of these ligands with the cyclophilins and FK506 binding proteins, respectively], inhibit the calcium/calmodulin-dependent phosphatase calcineurin (29–31) suggested routes to related drug discovery targets. This work highlighted the important role of this phosphatase in immune regulation and in other widespread calcium-activated signaling pathways, ranging from the recovery from cell cycle arrest by α-mating factor in yeast to the opening of an inward potassium channel involved in stomatal opening in plant guard cells (32). The broader concept of a small molecule acting through complexing large molecules recalls the recognition of antigen bound to histocompatibility molecules by the T-cell receptor (33).

Other organisms are good sources of agents that selectively block the action of coagulation proteases. Not surprisingly, these include leeches (34) and ticks (35). The plasminogen activator evolved by vampire bats appears to be even more fibrin-dependent (and, therefore, more clot selective) than that evolved by humans and used clinically (36, 37).

While the usefulness of anticoagulants and fibrinolytic agents to blood-sucking organisms appears obvious, it is less obvious what the selective advantage of opium alkaloid synthesis is to the poppy plant. Whatever its role in nature, morphine's interaction with the vertebrate central nervous system played an important role in the birth of molecular pharmacology. It led to the identification and isolation of human receptors that bind to opioids and to the identification of endogenous ligands for these receptors (38, 39). Some organisms provide not drugs or even leads but research tools (18, 40). For example, a screen can be designed to identify small molecules that block the binding of toxins from spiders, scorpions, or cone snails to ion channels. Potential uses for such small molecule receptor blockers range from limiting the extent of damage after a stroke (41) to treating asthma (42).

SELECTING SOURCES TO BE SCREENED FOR DRUGS AND LEADS

If organisms were selected for screening by focusing solely on taxa that had previously yielded interesting structures (43), the potential of as yet unexplored groups of organisms would be missed. Our screens span a broad range of diseases and, therefore, many molecular targets. To increase the chance of finding drugs or at least leads for a medicinal chemistry effort, we gather a diverse array of molecular structures for screens. Molecular diversity is sampled through our chemical collections and by collecting organisms from a broad range of sources. Fungi that live in complex environments and synthesize molecules that regulate the growth of neighboring insects, plants, bacteria, and/or other fungi are likely to be a potentially rich source of leads (44). A program established to maximize molecular diversity by screening microorganisms from a wide range of phylogenetic and ecological origins, such as endophytic (45) and coprophilic fungi (46), contributed to the discovery of a new class of metabolites, the zaragozic acids (Figure 2), which inhibit the first committed step in the isoprenoid pathway dedicated to sterol biosynthesis in fungi and mammals (48). Thus, zaragozic acids are broad-spectrum fungicides and also can lower mammalian cholesterol levels. Other interesting fungal metabolites identified by this approach include the antifungal viridofungans (49), selective nonpeptidyl inhibitors of the antitumor target Ras farnesyl-protein transferase (50), and an inhibitor of the HIV protease (51) (Figure 2).

Plants are another source of molecules with dramatic activity on both

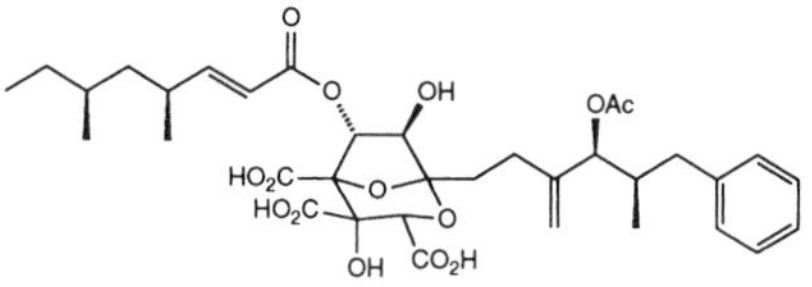

FIGURE 2 Examples of the variety of structures obtained in natural product screens. **I**, zaragozic acid A, is an inhibitor of mammalian and fungal sterol synthesis, obtained from fungi (48); **II**, L-696,474, is an inhibitor of the HIV protease, obtained from fungi (51); **III**, dehydrosoyasaponin I, is an agonist of the calcium-activated potassium channel, obtained from a medicinal plant (58); **IV**, tetrandrine, is an inhibitor of L-type calcium channels, obtained from a plant (78).

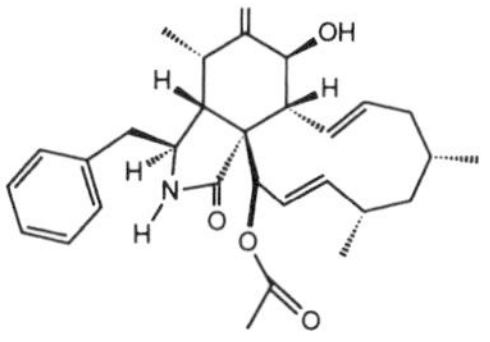

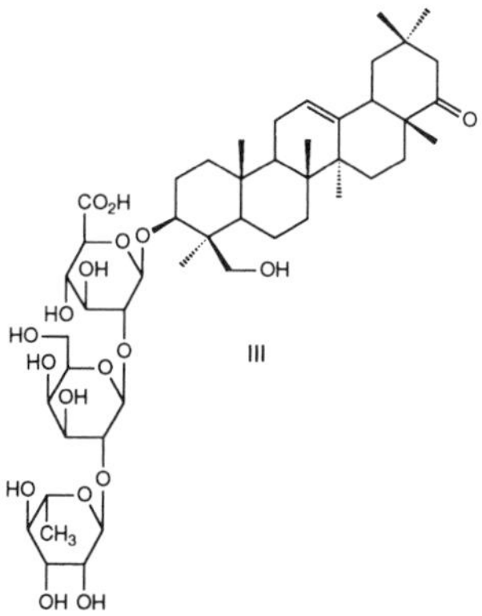

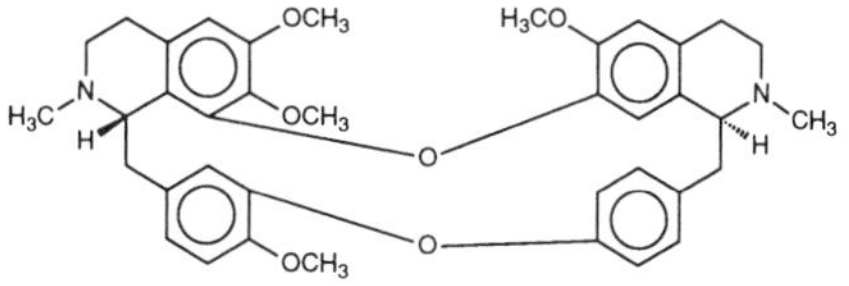

invertebrates and vertebrates. Some of these activities may have evolved under selective pressure from foraging animals. For example, plants that produce phytoecdysones can disrupt the development of insects that might devour them. The ecdysone mimic of *Cycas*, cycasterone, is more potent and less susceptible to metabolic inactivation than α-ecdysone itself (52). Bruchid beetles have been observed to avoid seeds of the legume *Mucuna* and may do so because of the high concentrations of L-dopa contained in these seeds (53). (However, L-dopa, used to treat Parkinson disease, was not originally discovered by following beetle behavior.)

A fungus that grows on wheat expends resources to produce ergot alkaloids, which affect vertebrate monoamine neurotransmission. Do ergot alkaloids have an internal plant/fungus purpose or have they evolved as compounds that protect the plant host and, thus, also the fungus from foraging vertebrates (54)? A plant growth hormone analogue, an insect antifeedant, and a tremorgenic compound that affects sheep and cattle are synthesized by ryegrass when it is infected with the endophyte *Acremonium loliae* (55). This may explain the observation that infected ryegrass is taller and more disease-resistant than uninfected ryegrass (56). However, while plants that resist foragers are potential sources of compounds that interact with human receptors and enzymes, many of these compounds are toxic agents, not drugs. Plants also can be selected for screening based on the advice of indigenous peoples (43, 57); for example, knowledge that *Desmodium adscendens* is used as a medicinal herb in Ghana for treatment of asthma provided a long-sought agonist of the maxi-K channel (58). This biochemical activity is consistent with this herb's use in other conditions associated with smooth muscle contraction (59).

Because compounds produced by one species may be modified by another species, knowledge of the natural history of an organism from which a lead was obtained may prove useful to medicinal chemists. For example, a male *Danaid* butterfly modifies *Senecio* alkaloids to synthesize pheromones (60); the female pine bark beetle *Dendoctonus brevicomus* and microorganisms in her gut alter the monoterpenes of the ponderosa pine (52). Thus, to assist medicinal chemists in their exploration of how modifications in the structure of the lead affect its activity, ecologists could suggest potential sources of structures derived from the lead. Understanding the biochemical basis of an ecological interaction has the potential to guide us to drugs (61). For example, the enzyme hydroxymethylglutaryl (HMG)-CoA reductase catalyzes a key step in sterol biosynthesis, the conversion of acetyl-CoA to mevalonate. As mevalonate has been shown to overcome catabolite repression of gibberellin synthesis (62), it seems likely that an inhibitor of HMG-CoA reductase plays a role in regulation of gibberellin synthesis. (Gibberellins, first isolated from the phytopathogenic fungus *Gibberella fujikuroi*, are plant growth hormones that stimulate stem elongation, induce flowering, and overcome seed dormancy.) Thus, if asked where to look for an inhibitor of HMG-CoA reductase, a chemical ecologist studying gibberellins might have suggested screening such phytopathogenic fungi. [The importance of HMG-CoA reductase inhibition in medicine is demonstrated by lovastatin (2, 63), discovered by random screening of fungi, developed as a drug, and now used by several million people in the United States alone to lower plasma cholesterol (64). Lovastatin may regulate sterol biosynthesis in the fungi that synthesize it or interfere with sterol biosynthesis in a fungal competitor, but its role in nature is not known.]

The leaf cutter ant *Atta cephalotes*, which cultivates a fungus as food, produces compounds that increase the growth of the fungal food and compounds that block the germination of undesired fungal spores, bacteria, and pollen (52, 65, 66). This ant avoids leaves of *Hymenaea courbaril*, which contain a terpenoid harmful to the fungus that the ant cultivates for food (67). Thus, collecting leaves avoided by an ant can point us to antifungal compounds.

The antiparasitic agent Ivermectin, which can prevent river blindness, was discovered in the microorganism *Streptomyces avermitilis* (68). *S. avermitilis* may synthesize a nematocide to kill nearby nematodes as a source of food (69), to eliminate a nematode that might otherwise eat the fungus or compete with it for food, or for a biologically unrelated purpose involving a target structurally related to Ivermectin's molecular target. An improved understanding of chemical ecology might allow us to predict where to find such valuable molecules.

Certain compounds appear in nature only when specific organisms interact. For example, the wild tobacco plant *Nicotiana sylvestris* increases its synthesis of alkaloids when under attack from larvae of *Manduca sexta* (70). Pathogens may alter plant gene expression (71) and trigger the synthesis of compounds that help resist pathogen attack (72, 73). While plants can respond to fungus-derived signals by synthesizing protective phytoalexins, fungi can respond by preventing phytoalexin accumulation or by detoxifying a specific phytoalexin by enzymatic conversion to a new structure (52, 73). Such a metabolite is missed by routine high throughput screens, which do not evaluate the plant and its fungus together.

In the laboratory, cells from the plant *Digitalis lanata* do not produce cardiac glycosides unless they differentiate (74). In the field, the concentration of codeine in the opium poppy varies with the hour of the day (75). A dramatic example of the influence of the environment on our ability to find potent compounds in an organism is seen with the poison dart frog *Phyllobates terribilis*. While a lethal dose of the voltage-dependent sodium channel agonist batrachotoxin can be harvested by rubbing the tip of a blow dart across the back of a field-caught specimen, batrachotoxin could not be detected in second generation terrarium-reared *P. terribilis*. Does batrachotoxin, or an essential precursor, come from the frog's diet, and/or is there an environmental trigger of its synthesis (76)? In either case, if the poison arrow frog had been studied only in the laboratory, we would be completely unaware of batrachotoxin. Similarly, if an organism that interacts with one we evaluate becomes extinct, we may miss finding potentially valuable molecules (72, 77).

INTEGRATING INFORMATION FROM MANY ORGANISMS TO GAIN INSIGHT FOR DRUG DESIGN

Natural product screening may capture a compound, such as an antibiotic, insecticide, or HMG-CoA reductase inhibitor, designed for the very purpose of interacting with the target of the screen. Since diverse organisms, from a quetzal to a Sequoia or from a fungus to human beings, share a great deal of biochemical architecture, it is not surprising that small molecules from species as distant as plants, fungi, and bacteria can interact with macromolecules in our bodies. Natural product screening allows us to sift through the extraordinary diversity that overlays these common molecular features. It may be that an interaction is fortuitous, given the number and diversity of natural products examined in pharmaceutical industry screens. However, a compound, discovered in a screen, that interacts with a human protein may have evolved to interact with a protein bearing a homologous domain, active site, or regulatory region in the organism within which it is made (or in a neighboring organism). For example, might tetrandrine (found in a Chinese medicinal herb, *Stepania tetrandra*, used to treat angina and hypertension), which blocks human L-type calcium channels (78), have evolved to regulate plant calcium channels (79)? Information about related proteins from different organisms can help us to understand how a target protein's function emerges from its structure. Amino acids conserved in related receptors produced by different species can guide us to residues important, for example, in activating the receptor (80). Sequences of variant receptors or enzymes with unique pharmacology point to key residues of a target important for drug recognition. Thus, the Monarch butterfly *Danaus plexippus* is able to feed on plants rich in cardiac glycosides that are toxic for most animals. The substitution His-122 → Asn in its Na^+/K^+-ATPase has been proposed to explain the butterfly's decreased sensitivity to the cardiotonic agent ouabain (81).

As masses of DNA sequences accumulate for protein family members from a wide range of organisms, we can learn to recognize binding motifs that define active and regulatory sites. A study of homologous proteins from diverse organisms can provide a new level of insight into their activity. For example, the exact distance between the leucine-zipper and the basic region proposed as a DNA contact surface was noted to be conserved in DNA binding proteins derived from a range of species including plants, fungi, and mammals. This observation led to a helical model for the binding of this family of transcription regulatory proteins to DNA and to an understanding of how dimers of such proteins could specifically recognize directly abutted dyad-symmetric DNA sequences (82) (Figure 3). A new science, bioinformatics, is evolving to extract information, with increasing levels of abstraction, from a burgeoning list of DNA sequences (84). Even otherwise unrelated proteins may have related stretches of amino acid sequence (motifs) (85, 86) and/or similar backbone structure (87) at the binding site. While most computer searches for such motifs focus on the primary sequence of the encoded protein, it

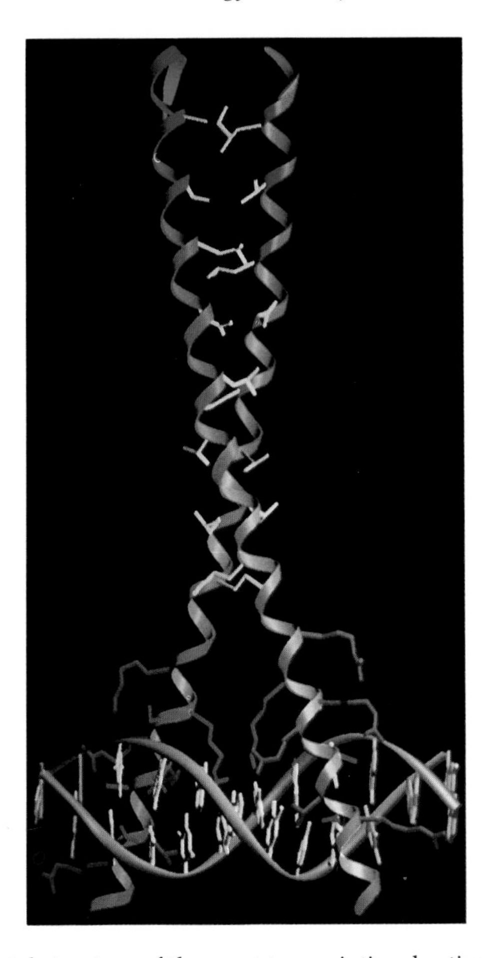

FIGURE 3 Crystal structure of the yeast transcriptional activator GCN4, bound to DNA (83). This structure confirms the prediction that had been based on the alignment of 11 sequences from a range of species that the conserved leucines (highlighted in yellow) would be oriented toward each other along the faces of neighboring helices. As predicted, the helices cross the DNA in the major groove. Surprisingly, the GCN4 residues making specific DNA base contacts (highlighted in red) are conserved among proteins with different DNA binding specificities, whereas the amino acids contacting the DNA backbone in the GCN4 structure (highlighted in blue) are less conserved in this protein family. Thus, while predictions made based upon linear amino acid sequences from 11 species were confirmed, we await three-dimensional structures from diverse sources to elucidate how variant amino acids alter the orientation of the conserved amino acids to effect base-sequence-specific recognition (figure by Tom Ellenberger).

is beginning to be feasible to search for homology in the three-dimensional structures that linear sequences can form (84, 86, 89). While certain of these homologous regions may interact with the same compound [e.g., ATP (85)], others may interact with structurally homologous but distinct ligands such as individual members of a protein family.

Comparing the sequences of two families of macromolecules that interact, such as DNA sequences and the proteins that recognize them, helps us to perceive the molecular basis of specificity (90). Variant amino acids that provide an individual sequence's functional specificity are intermixed with consensus amino acids, which provide a motif's structural framework. A compound (whether discovered in a screen or designed by medicinal chemists) that binds to a given sequence in one protein may provide a lead for a compound that can bind to that motif in other proteins. Such information combined with advances in analytical (91) and molecular (92) modeling methodology dramatically improves our ability to visualize how recognition is framed in three dimensions. This insight can reveal a rational route to improved potency and specificity. Because members of protein families have related three-dimensional structures, small molecules that interact with the members of a given protein family also may have common underlying three-dimensional structures, which may be viewed as "scaffolds." Specificity of a small molecule for an individual protein family member may be overlayed on a common scaffold by groups that provide specific binding interactions (hydrophobic bonds, salt bridges, hydrogen bonds, etc.). For example, scaffolds designed to interact with an important family of signal transducing molecules, G-protein-linked receptors, provide an approach to a novel set of potentially selective ligands (47) (Figure 4). Even when the biological ligand of our target is another protein or peptide, we are beginning to be able to select or design scaffold structures that direct interactions to the right places in space (93, 94). Favored scaffolds upon which to introduce specificity-defining binding interactions include classes of molecules known to have generally good bioavailability, such as steroids and benzodiazepines (95, 96) (much as the steroid nucleus itself serves, in nature, as a scaffold on which different substituents are responsible for selective interaction with individual members of the steroid receptor protein family); other scaffolds may be found through screening natural products.

EFFICIENT ROUTES TO THE DISCOVERY OF USEFUL COMPOUNDS

Life generates possibilities by providing opportunities for randomness, upon which selection acts. However, untethered randomness is not the most efficient route to success in a competitive world. Organisms have

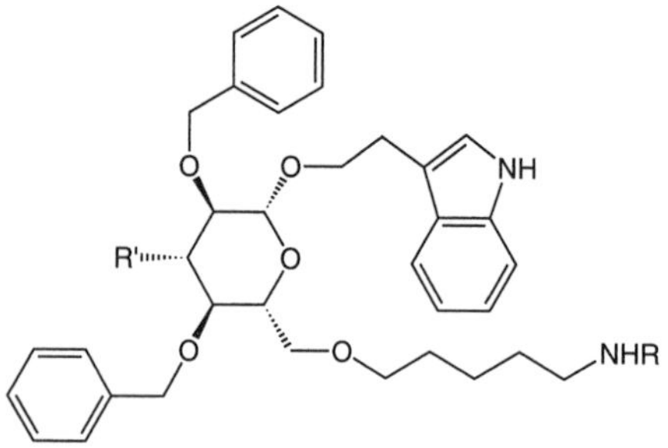

R	R'	Somatostatin	Substance P	β_2
H	H	8.4	0.50	500
$CH_3C(=O)$	OCH_2–Ph	>100	0.06	>500
H	OCH_2–Ph 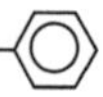	15	0.15	90

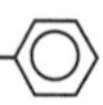

FIGURE 4 Scaffold that interacts selectively with different members of a protein family. These β-D-glucose derivatives (47) can compete with each G-protein-linked receptor's natural ligand for binding, whether the natural ligand is a peptide or a catecholamine. (Note that the natural ligands do not compete with each other for binding.) By varying the substituents, the ability to discriminate between members of a protein family can be overlayed on a common scaffold. β_2, β_2 receptor.

evolved a variety of efficient mechanisms to acquire new traits, such as passing around antibiotic resistance genes (97) or copying and modifying useful structures (98, 99). In the development of an immune response against a pathogen, DNA sequences encoding variable regions, which provide the specificity for the pathogen, are moved next to constant regions, which provide for different effector functions (such as crossing the placenta, binding to phagocytes, or degranulating mast cells) (100). While this occurs during the life of an individual, it appears that a similar process occurs during evolution (98, 99, 101, 102). This assembling of DNA sequences from different sources generates patchwork genes, such as that encoding the low density lipoprotein receptor (103). Similarly, new serine proteases appear to have been created by duplication of DNA encoding a functional serine protease domain, followed by mutation (104) and attachment of modules with varied functions (105). Thus, long before the industrial revolution, nature discovered the efficiency of interchangeable parts (106). Mechanisms that direct genetic variation away from a useful framework and toward sites where such variation can generate novel functions rather than destroy useful ones are likely to have been evolutionarily selected (107). For example, after duplication of a serine protease gene, a functional serine protease that is specific for a new substrate arises through mutations that alter amino acids located in those regions of the protease that determine substrate specificity (108). The most efficient route to new active serine proteases would tend to avoid mutations likely to inactivate the enzyme, such as those in the active site or framework of the three-dimensional serine protease structure. Exploration could be directed to specific stretches of DNA, for example, by modulating the fidelity of replication (109) and employing the degeneracy of the genetic code to generate sequence-specific effects on DNA structure (107). Another suggestion of directed variation involves cone snails, which move at a snail's pace yet prey on fish, which move quickly. The snail must paralyze the fish instantly. To meet this challenge, cone snails have evolved a family of peptides (110) with a conserved ion-channel active framework. It has been proposed that the genus *Conus* uses a strategy involving cassette mutagenesis to conserve the ion channel active scaffold while generating specificity for diverse channels (111). A recombination-based mechanism for targeted diversity also has been proposed for histocompatibility antigens (112). Somatic mutation appears to be directed to the variable (antigen binding) region of immunoglobulin molecules (113). The better we understand how specificity is overlaid on conserved frameworks in biochemical targets, the more efficient our design of potent and specific drugs will be.

Once we obtain a structure that interacts with a target (whether through computerized searches and assays of a compound collection or through natural product screening), variants of this structure are synthesized to optimize potency and specificity. It is becoming possible to explore variations of some leads rapidly through combinatorial synthesis (114) in which, typically, solid-phase synthesis (115) is adapted to perform a mixed sequence of reactions simultaneously, generating hundreds or hundreds of thousands of products in a single synthesis. Such a combinatorial approach to drug discovery is dependent upon both the availability of a rapid screening method to select active molecules and a rapid method of defining their structures and potencies. One source of creativity for combinatorial chemistry is the set of synthons used in combination. Enzymes obtained from microorganisms are likely to be enlisted to help us perform challenging synthetic steps, increasing the diversity of structures that are accessible to combinatorial chemistry.

In the absence of any information about the structure of a target, combinatorial synthesis can be used to explore a set of diverse scaffolds that direct potential binding interactions to different angles and distances from each other. The structure–activity relationship that emerges from such a study helps to define a three-dimensional array that provides successful ligands for the target. A scaffold that fits the target may exist in nature fortuitously, having a different function but just the right shape to interact with our target, and/or may have evolved when nature confronted the same structural problem as that before us. Scaffolds previously discovered to interact with a particular receptor family or motif can be used as a starting point for combinatorial chemistry. As our ability to carry out combinatorial chemistry evolves, libraries of molecules will emerge that include compounds selective for each variant of a protein motif. Whether a useful compound emerges from variation and natural selection or through combinatorial chemistry and laboratory screening, our ability to capture compounds selective for a target protein, combined with increased understanding of the structure and function of conserved and variant regions of protein targets, is enabling us to evolve more efficient strategies for discovering useful molecules.

LEARNING TO ASK BETTER QUESTIONS

A key step in the drug discovery process is the identification of a target and demonstration that its activity affects the disease process. We can learn to identify key points of biological regulation by studying the fruits of evolution, such as cone snail toxins, vampire bat fibrinolytics, viruses that interfere with cell death, fungal and bacterial immunosuppressants, and human genetic variation. We can be inspired in our search for novel drugs by the diverse interactions revealed by research in chemical ecology.

Evolution is a process of generating possibilities and selecting among them. The value of a new biological activity is not an absolute and not an intrinsic property of the new structure; rather, value depends on the possibilities that the new property generates in the context of other compounds or organisms in the environment in which the activity emerged (116). Life is challenged by interactions to evolve. Natural selection may favor not only those organisms that produce molecules optimal for their purposes but also those lineages that have evolved the most efficient processes by which to generate optimal pathways and molecules. Molecular mechanisms have evolved that enlist and modify useful structures for new purposes. We are now beginning to search for efficient strategies by which to discover new molecules to protect us against faltering biochemical regulatory mechanisms and evolving infectious agents. The more information we have about the structure of a drug discovery target and its relationship to other proteins, the more efficiently we can optimize potency and selectivity in our drug discovery process. Drug design efforts can begin with a structure that was previously designed or one that was discovered through natural product screening. Combinatorial chemistry will allow us to add rapidly to the rich variety of biologically active molecules that has emerged from molecular evolution.

In the use of natural product screening for drug discovery, our descendants may better understand why a given activity and/or structure can be found in this molecule of this cell of this organism at this stage in its life cycle. A higher level of understanding should emerge from studying many compounds and many organisms in many complex settings. As more species are lost, our biochemical reference library becomes poorer. As we study enough threads and their connections, we can begin to see patterns. These patterns form a framework into which we can place observations and identify gaps in our knowledge. The loss of the shamans' knowledge of the medicinal uses of plants, as their cultures disappear, has been compared to the burning of the library at Alexandria (117). But for those of us who are beginning to understand how to read the molecules within living things, the loss of biological diversity itself is also the loss of a library, a library that contains answers to questions we have not yet learned to ask.

SUMMARY

Biological diversity reflects an underlying molecular diversity. The molecules found in nature may be regarded as solutions to challenges that have been confronted and overcome during molecular evolution. As our understanding of these solutions deepens, the efficiency with which we can discover and/or design new treatments for human disease grows. Nature assists our drug discovery efforts in a variety of ways. Some compounds synthesized by microorganisms and plants are used directly as drugs. Human genetic variations that predispose to (or protect against) certain diseases may point to important drug targets. Organisms that manipulate molecules within us to their benefit also may help us to recognize key biochemical control points. Drug design efforts are expedited by knowledge of the biochemistry of a target. To supplement this knowledge, we screen compounds from sources selected to maximize molecular diversity. Organisms known to manipulate biochemical pathways of other organisms can be sources of particular interest. By using high throughput assays, pharmaceutical companies can rapidly scan the contents of tens of thousands of extracts of microorganisms, plants, and insects. A screen may be designed to search for compounds that affect the activity of an individual targeted human receptor, enzyme, or ion channel, or the screen might be designed to capture compounds that affect any step in a targeted metabolic or biochemical signaling pathway.

While a natural product discovered by such a screen will itself only rarely become a drug (its potency, selectivity, bioavailability, and/or stability may be inadequate), it may suggest a type of structure that would interact with the target, serving as a point of departure for a medicinal chemistry effort—i.e., it may be a "lead." It is still beyond our capability to design, routinely, such lead structures, based simply upon knowledge of the structure of our target. However, if a drug discovery target contains regions of structure homologous to that in other proteins, structures known to interact with those proteins may prove useful as leads for a medicinal chemistry effort. The specificity of a lead for a target may be optimized by directing structural variation to specificity-determining sites and away from those sites required for interaction with conserved features of the targeted protein structure. Strategies that facilitate recognition and exploration of sites at which variation is most likely to generate a novel function increase the efficiency with which useful molecules can be created.

I am grateful to Thomas Eisner and Jerrold Meinwald for organizing this thought-provoking Colloquium; to Gerald Bills, Ralph Hirschmann, Joel Huff, Chris Sander, Catherine Strader, Gary Stoecker, and especially, Jerrold Meinwald for their very helpful comments on earlier versions of this manuscript; and to Marianne Belcaro for assistance with its preparation.

REFERENCES

1. Merck & Co. (1993) *Annual Report* (Merck & Co., Rahway, NJ).
2. Alberts, A. W., Chen, J., Kuron, G., Hunt, V., Huff, J., *et al.* (1980) *Proc. Natl. Acad. Sci. USA* **77,** 3957–3961.
3. Kahan, J. S., Kahan, F. M., Goegelman, R., Currie, S. A., Jackson, M., Stapley, E. O., Miller, T. W., Miller, A. K., Hendlin, D., Mochales, S., Hernandez, S., Woodruff, H. B. & Bimbaum, J. (1978) *J. Antibiot.* **32,** 1–12.
4. Ferreira, S. H. (1965) *Br. J. Pharmacol.* **24,** 163–169.
5. Ondetti, M. A., Rubin, B. & Cushman, D. W. (1977) *Science* **196,** 441–444.
6. Patchett, A. A., Harris, E., Tristam, E. W., Wyvratt, M., Wu, M. T., *et al.* (1980) *Nature (London)* **288,** 280–283.
7. Imperato-McGinley, J., Guerrero, L., Gautier, T., & Petersen, R. (1974) *Science* **186,** 1213–1215.
8. Kohl, N. E., Emini, E. A., Schleif, W. A., Davis, L. J., Heimbach, J. C., Dixon, R. A. F., Scolnick, E. M. & Sigal, I. S. (1988) *Proc. Natl. Acad. Sci. USA* **85,** 4686–4690.
9. Huff, J. R. (1991) *J. Med. Chem.* **34,** 2305–2314.
10. Wiley, R. A. & Rich, D. H. (1993) *Med. Res. Rev.* **13,** 327–384.
11. Lingham, R. B., Hsu, A., Silverman, K. C., Bills, G. F., Dombrowski, A., Goldman, M. E., Darke, P. L., Huang, L., Koch, G., Ondeyka J. G. & Goetz, M. A. (1992) *J. Antibiot.* **45,** 686–691.
12. Kashman, Y., Gustafson, K. R., Fuller, R. W., Cardellina, J. H., McMahon, J. B.,

Currens, M. J., Buckheit, R. W., Jr., Hughes, S. H., Cragg, G. M. & Boyd, M. R. (1992) *J. Med. Chem.* **35,** 2735–2743.

13. Lam, P. Y. S., Jadhav, P. K., Eyermann, C. J., Hodge, C. N., Ru, Y., Bacheler, L. T., Meek, J. L., Otto, M. J., Rayner, M. M., Wong, Y. N., Chang, C., Weber, P. C., Jackson, D. A., Sharpe, T. R. & Erickson-Viitanen, S. (1994) *Science* **263,** 380–384.
14. Thompson, W. J., Fitzgerald, P. M. D., Holloway, M. K., Emini, E. A., Darke, P. L., McKeever, B. M., Schleif, W. A., Quintero, J. C., Zugay, J. A., Tucker, T. J., Schwering, J. E., Homnick, C. F., Nunberg, J., Springer, J. P. & Huff, J. R. (1992) *J. Med. Chem.* **35,** 1685–1701.
15. Stinson, S. (1994) *C & E News* **5/16/94,** 6–7.
16. Vacca, J. P., Dorsey, B. D., Schleif, W. A., Levin, R. B., McDaniel, S. L., *et al.* (1994) *Proc. Natl. Acad. Sci. USA* **91,** 4096–4100.
17. Gadebusch, H. H., Stapley, E. O. & Zimmerman, S. B. (1992) *Crit. Rev. Biotechnol.* **12,** 225–243.
18. Monaghan, R. L. & Tkacz, J. S. (1990) *Annu. Rev. Microbiol.* **44,** 271–301.
19. Tanzi, R., Gaston, S., Bush, A., Romano, D., Pettingell, W., Peppercorn, J., Paradis, M., Gurubhagavatula, S., Jenkins, B. & Wasco, W. (1994) *Genetica* **91,** 255–263.
20. Watson, S. & Girdlestone, D., eds. (1994) *1994 Receptor and Ion Channel Nomenclature Supplement* (Elsevier, London).
21. Torphy, T. J. & Undem, B. J. (1991) *Throax* **46,** 512–523.
22. Busse, W. W. & Sosman, J. (1976) *Science* **194,** 737–738.
23. Gooding, L. R. (1992) *Cell* **71,** 5–7.
24. Kierszenbaum, F. (1994) *Parasitic Infections and the Immune System* (Academic, San Diego), pp. 163–164.
25. Bliska, J. B., Galan, J. E. & Falkow, S. (1993) *Cell* **73,** 903–920.
26. Henderson, S., Rowe, M., Gregory, C., Croom-Carter, D., Wang, F., Longnecker, R., Kieff, E. & Rickinson, A. (1991) *Cell* **65,** 1107–1115.
27. Crook, N. E., Clem, R. J. & Miller, L. K. (1993) *J. Virol.* **67,** 2168–2174.
28. Zychlinsky, A., Kenny, B., Menard, R., Prevost, M., Holland, I. B. & Sansonetti, P. J. (1994) *Mol. Microbiol.* **11,** 619–627.
29. McKeon, F. (1991) *Cell* **66,** 823–826.
30. Liu, J., Farmer, J. D., Lane, W. S., Friedman, J., Weissman, I. & Schreiber, S. L. (1991) *Cell* **66,** 807–815.
31. Clardy, J. (1995) *Proc. Natl. Acad. Sci. USA* **92,** 56–61.
32. Poole, R. J. (1993) *Proc. Natl. Acad. Sci.* **90,** 3125–3126.
33. Jorgensen, J. L., Reay, P. A., Ehrich, E. W. & Davis, M. M. (1992) *Annu. Rev. Immunol.* **10,** 835–873.
34. Kaiser, B., (1992) *Thromb. Haemostasis* **17,** 130–136.
35. Mellott, M. J., Stranieri, M. T., Sitko, G. R., Lynch J. J., Jr., & Vlasuk, G. P. (1993) *Fibrinolysis* **7,** 195–202.
36. Gardell, S. J., Duong, L. T., Diehl, R. E., York, J. D., Hare, T. R., Register, R. B., Jacobs, J. J., Dixon, R. A. F. & Friedman, P. A. (1989) *J. Biol. Chem.* **264,** 17947–17952.
37. Pennica, D., Holmes, W. E., Kohr, W. J., Harkins, R. N., Vehar, G. A., Ward, C. A., Bennett, W. F., Yelverton, E., Seeburg, P. H., Heyneker, H. L., Goeddel, D. V. & Collen, D. (1983) *Nature (London)* **301,** 214–221.
38. Pert, C. B. & Snyder, S. H. (1973) *Science* **179,** 1011–1014.
39. Kosterlitz, H. W. & Hughes, J. (1977) *Br. J. Psychiatry* **130,** 298–304.

40. Jones, T. R., Charette, L., Garcia, M. L. & Kaczorowski, G. J. (1990) *J. Pharmacol. Exp. Ther.* **255,** 697–705.
41. Valentino, K., Newcomb, R., Gadbois, T., Singh, T., Bowersox, S., Bitner, S., Justice, A., Yamashiro, D., Hoffman, B. B., Ciaranello, R., Miljanich, G. & Ramachandran, J. (1993) *Proc. Natl. Acad. Sci. USA* **90,** 7894–7897.
42. Huang, J., Garcia, M. L., Reuben, J. P. & Kaczorowski, G. J. (1993) *Eur. J. Pharmacol.* **235,** 37–43.
43. Balick, M. J. (1990) in *Bioactive Compounds from Plants,* Ciba Foundation Symposium 154, eds. Chadwick, D. J. & Marsh, J. (Wiley, Chichester, U.K.), pp. 22–39.
44. Nisbet, L. J., Fox, F. M. & Hawksworth, D. L. (1991) *The Biodiversity of Microorganisms and Invertebrates: Its Role in Sustainable Agriculture* (CAB Int., Wallingford, U.K.), pp. 229–244.
45. Bills, G. F. (1994) in *Systematics, Ecology and Evolution of Endophytic Fungi in Grasses and Woody Plants,* ed. Redlin, S. C. (APS Press, St. Paul), in press.
46. Bills, G. F. & Polishook, J. D. (1993) *Nova Hedwigia Z. Kryptogramenkd.* **57,** 195–206.
47. Hirschmann, R., Nicolau, K. C., Pietranico, S., Leahy, E. M., Salvino, J., *et al.* (1993) *J. Am. Chem. Soc.* **115,** 12550–12568.
48. Bills, G. F., Palaez, F., Polishook, J. D., Diez-Matas, M. T., Harris, G. H., Clapp, W. H., Dufresne, C., Byrne, K. M., Nallin-Omstead, M., Jenkins, R. G., Mojena, M., Huang, L. & Bergstrom, J. D. (1994) *Mycol. Res.* **98,** 733–739.
49. Harris, G. H., Jones, E. T. T., Meinz, M. S., Nallin-Omstead, M., Helms, G. L., Bills, G. F., Zink, D. & Wilson, K. E. (1993) *Tetrahedron Lett.* **34,** 5235–5238.
50. Singh, S. B., Zink, D. L., Liesch, J. M., Goetz, M. A., Jenkins, R. G., Nallin-Omstead, M., Silverman, K. C., Bills, G. F., Mosley, R. T., Gibbs, J. B., Albers-Schonberg, G. & Lingham, R. B. (1993) *Tetrahedron* **49,** 5917–5926.
51. Dombrowski, A. W., Bills, G. F., Sabnis, G., Koupal, L. R., Meyer R., Ondeyka, J. G., Giacobbe, R. A., Monaghan, R. L. & Lingham, R. B. (1992) *J. Antibiot.* **45,** 671–678.
52. Harborne, J. B. (1988) *Introduction to Ecological Biochemistry* (Academic, New York), 3rd Ed.
53. Janzen, D. (1969) *Evolution* **23,** 1–27.
54. Wink, M. (1993) in *The Alkaloids: Chemistry and Pharmacology,* ed. Cordell, G.A. (Academic, New York), pp. 1–104.
55. Rowan, D. D., Hunt, M. B. & Gaynor, D. L. (1986) *J. Chem. Soc. Chem. Commun.* 935–936.
56. Clay, K. (1991) in *Microbial Mediation of Plant-Herbivore Interactions,* eds. Barbosa, P., Krischik, V. A. & Jones, C. G. (Wiley, New York), pp. 199–226.
57. Cox, P. A. (1990) in *Bioactive Compounds from Plants,* Ciba Foundation Symposium 154, eds. Chadwick, D. J. & Marsh, J. (Wiley, Chichester, U.K.), p. 40.
58. McManus, O. B., Harris, G. H., Giangiacomo, K. M., Feigenbaum, P., Reuben, J. P., Addy, M. E., Burka, J. F., Kaczorowski, G. J. & Garcia, M. L. (1993) *Biochemistry* **32,** 6128–6133.
59. Arvigo, R. & Balick, M. (1993) *Rainforest Remedies* (Lotus, Twin Lakes, WI).
60. Eisner, T. & Meinwald, J. (1995) *Proc. Natl. Acad. Sci. USA* **92,** 50–55.
61. Horn, W. S., Smith, J. L., Bills, G. F., Raghoobar, S. L., Helms, G. L., Kurtz, M. B., Marrinan, J. A., Frommer, B. R., Thornton, R. A. & Mandala, S. M. (1992) *J. Antibiot.* **45,** 1692–1696.
62. Bruckner, B. (1992) *Secondary Metabolites: Their Function and Evolution,* Ciba

Foundation Symposium 171, eds. Chadwick D. J. & Whelan J. (Wiley, Chichester, U.K.), pp. 129–143.
63. Slater, E. E. & MacDonald, J. S. (1988) *Drugs* **36S3,** 7–82.
64. Anonymous (1993) *Stat. Bull.*, 10–17.
65. Iwanami, Y. (1978) *Protoplasma* **95,** 267–271.
66. Cherrett, J. M., Powell, R. J. & Stradling, D. J. (1988) in *Coevolution of Fungi with Plants and Animals*, eds. Pirozynski, K.A. & Hawksworth, D.L. (Academic, New York), pp. 93–120.
67. Hubbell, S. P., Wiemer, D. F. & Adejare, A. (1983) *Oecologia (Berlin)* **60,** 321–327.
68. Burg, R. W., Miller, B. M., Baker, E. E., Birnbaum, J., Currie, S. A., Hartman, R., Kong, Y., Monaghan, R. L., Olson, G., Putter, I., Tunac, J. B., Wallick, H., Stapley, E. O., Oiwa, R. & Omura, S. (1979) *Antimicrob. Agents Chemother.* **15,** 361–367.
69. Giuma, A. Y. & Cooke, R. C. (1971) *Trans. Br. Mycol. Soc.* **56,** 89–94.
70. Harborne, J. B. (1990) *Bioactive Compounds from Plants*, Ciba Foundation Symposium 154, eds. Chadwick, D. J. & Marsh, J. (Wiley, Chichester, U.K.), pp. 126–139.
71. Collinge, D. B. & Slusarenko, A. J. (1987) *Plant Mol. Biol.* **9,** 389–410.
72. Gaffney, T., Friedrich, L., Vernooij, B., Negrotto, D., Nye, G., Uknes, S., Ward, E., Kessmann, H. & Ryals, J. (1993) *Science* **261,** 754–756.
73. Barz, W., Bless, W., Borger-Papendorf, G., Gunia, W., Mackenbrock, U., Meier, D., Otto, Ch. & Super, E. (1990) *Bioactive Compounds from Plants*, Ciba Foundation Symposium 154, eds. Chadwick, D. J. & Marsh, J. (Wiley, Chichester, U.K.), pp. 140–156.
74. Luckner, M. & Diettrich, B. (1989) in *Primary and Secondary Metabolism of Plant Cell Cultures*, ed. Kurz, W. G. W. (Springer, Berlin), pp. 117–124.
75. Grindley, J. (1993) *Scrip Mag.* **12/93,** 30–33.
76. Daly, J. W. (1995) *Proc. Natl. Acad. Sci. USA* **92,** 9–13.
77. Enyedi, A. J., Yalpani, N., Silverman, P. & Raskin, I. (1992) *Cell* **70,** 879–886.
78. King, V. F., Garcia, M. L., Himmel, D., Reuben, J. P., Lam, Y. T., Pan, J., Han, G. & Kaczorowski, G. J. (1988) *J. Biol. Chem.* **263,** 2238–2244.
79. Graziana, A., Fosset, M., Ranjeva, R., Hetherington, A. M. & Lazdunski, M. (1988) *Biochemistry* **27,** 764–768.
80. Strader, C. D., Sigal, I. S., Register, R. B., Candelore, M. R., Rands, E. & Dixon, R. A. F. (1987) *Proc. Natl. Acad. Sci. USA* **84,** 4384–4388.
81. Holzinger, F., Frick, C. & Wink, M. (1992) *FEBS Lett.* **314,** 477–480.
82. Vinson, C. R., Sigler, P. B. & McKnight, S. L. (1989) *Science* **246,** 911–916.
83. Ellenberger, T. E., Brandl, C. J., Struhl, K. & Harrison, S. C. (1992) *Cell* **71,** 1223–1237.
84. Taylor, W. R. (1992) *Patterns in Protein Sequence and Structure*, Springer Series in Biophysics (Springer, Berlin), Vol. 7.
85. Bairoch, A. (1992) *Nucleic Acids Res.* **20** (Suppl.), 2013–2018.
86. Johnson, M. S., Overington J. P. & Blundell, T. L. (1993) *J. Mol. Biol.* **231,** 735–752.
87. Rizo, J. & Gierasch, L. M. (1992) *Annu. Rev. Biochem.* **61,** 387–418.
88. Sheridan, R. P. & Venkataraghavan, R. (1992) *Proteins* **14,** 16–28.
89. Ouzounis, C., Sander, C., Scharf, M. & Schneider, R. (1993) *J. Mol. Biol.* **232,** 805–825.
90. Desjarlais, J. R. & Berg, J. M. (1993) *Proc. Natl. Acad. Sci. USA* **90,** 2256–2260.

91. Roberts, G. C. K. (1993) *NMR of Macromolecules* (Oxford Univ. Press, New York).
92. Cohen, F. E. & Moos, W., eds. (1993) *Perspect. Drug Discovery Des.* **1,** 391–417.
93. Olson, G. L., Bolin, D. R., Bonner, M. P., Bos, M., Cook, C. M., Fry, D. C., Braves, B. J., Hatada, M., Hill, D. E., Kahn, M., Madison, V. S., Rusiecki, V. K., Sarabu, R., Sepinwall, J., Vincent, G. P. & Voss, M. E. (1993) *J. Med. Chem.* **36,** 3039–3049.
94. Smith, A. B., Hirschmann, R., Pasternak, A., Akaishi, R., Guzman, M. C., Jones, D. R., Keenan, T. P., Sprengeler, P. A., Darke, P. L., Emini, A., Holloway, M. K. & Schleif, W. A. (1994) *J. Med. Chem.* **37,** 215–218.
95. Hirschmann, R., Sprengeler, P.A., Kawasaki, T., Leahy, J. W., Shakespeare, W. C. & Smith, A. B., III (1992) *J. Am. Chem. Soc.* **114,** 9699–9701.
96. Bunin, B. A., Plunkett, M. J. & Ellman, J. A. (1994) *Proc. Natl. Acad. Sci. USA* **91,** 4708–4712.
97. Zahner, H., Anke, H. & Anke, T. (1983) in *Secondary Metabolism and Differentiation in Fungi,* eds. Bennett, J. W. & Ciegler, A. (Dekker, New York), pp. 153–171.
98. Doolittle, R. & Bork, P. (1993) *Sci. Am.* **269,** 50–56.
99. Caporale, L. H. (1973) Ph.D. Thesis (Univ. of California, Berkeley).
100. Hozumi, N. & Tonegawa, S. (1976) *Proc. Natl. Acad. Sci. USA* **73,** 3628–3632.
101. Rossman, M. G. (1981) *Philos. Trans. R. Soc. London B.* **293,** 191–203.
102. Gilbert, W. (1978) *Nature (London)* **271,** 501.
103. Sudhof, T. C., Goldstein, J. L., Brown, M. S. & Russell, D. W. (1985) *Science* **228,** 815–822.
104. Patthy, L. (1993) *Methods Enzymol.* **222,** 10–28.
105. Irwin, D. M., Robertson, K. A. & MacGillivray, R. T. A. (1988) *J. Mol. Biol.* **200,** 31–45.
106. Hays, S. P. (1957) *The Response to Industrialism 1885–1914* (Univ. of Chicago Press, Chicago).
107. Caporale, L. H. (1984) *Mol. Cell. Biochem.* **64,** 5–13.
108. Graham, L. D., Haggett, K. D., Jennings, P. A., De Brocque, D. S. & Whittaker, R. G. (1993) *Biochemistry* **32,** 6250–6258.
109. Kunkel, T. A. (1992) *J. Biol. Chem.* **267,** 18521.
110. Woodward, S. R., Cruz, L. J., Olivera, B. M. & Hillyard, D. R. (1990) *EMBO J.* **9,** 1015–1020.
111. Myers, R. A., Cruz, L. J., Rivier, J. E. & Olivera, B. M. (1993) *Chem. Rev.* **93,** 1923–1930.
112. Pease, L. R., Schulze, D. H., Pfaffenbach, G. M. & Nathenson, S. G. (1983) *Proc. Natl. Acad. Sci. USA* **80,** 242–246.
113. Pech, M., Höchtl, J., Schnell, H. & Zachau, H. (1981) *Nature* (London) **291,** 668–670.
114. Gordon, E. M., Barrett, R. W., Dower, W. J., Fodor, S. P. A. & Gallop, M. A. (1994) *J. Med. Chem.* **37,** 1385–1401.
115. Merrifield, B., (1986) *Science* **232,** 341–347.
116. Eiseley, L., (1959) *The Immense Journey* (Vintage, New York).
117. Linden, E. (1991) *Time* **9/3/91,** 46–55.

Abbreviations

9-HPEPE	9-hydroperoxyicosa-(5*Z*, 7*E*, 11*Z*, 14*Z*, 17*Z*)-pentaenoic acid
AL	antennal lobe
CNS	central nervous system
CS diet	*Crotalaria spectabilis* seed-supplemented diet
CsA	cyclosporin A
CyP	cyclophilin
e.e.	enantiomeric excess
FE	female equivalent
FKBP	FK506-binding protein
GABA	γ-aminobutyric acid
HD	hydroxydanaidal
HIV	human immunodeficiency virus
HMG	hydroxymethylglutaryl
hpp	hours postparasitization
IL-2	interleukin 2
IPSP	inhibitory postsynaptic potential
LAL	lateral accessory lobe
LLE	long-lasting excitation
LN	local interneuron
MGC	macroglomerular complex

MHC major histocompatibility complex

NF-AT nuclear factor for activated T cells

ORC olfactory receptor cell

PA pyrrolizidine alkaloid
PB diet pinto bean-based diet
Ph-I==O iodosylbenzene
PN projection neuron
PPIase peptidylprolyl isomerase

TCR T-cell receptor
TPPMn tetraphenylporphyrinatomanganese

Index

A

D

E

F

G

H

I

K

L

M

N

Q

R